# Lecture Notes in Mathematics



Edited by A. Dold and B. Eckmann

Subseries: Instituto de Matemática Pura e Aplicada, Rio de Janeiro

Adviser: C. Camacho

1207

Pierre H. Bérard

# Spectral Geometry: Direct and Inverse Problems

With an Appendix by G. Besson

Springer-Verlag

Berlin Heidelberg New York London Paris Tokyo

**Author**

Pierre H. Bérard
Département de Mathématiques, Université de Savoie
B.P. 1104, 73011 Chambéry Cedex, France

This book is being published in a parallel edition by the Instituto de Matemática Pura e Aplicada, Rio de Janeiro as volume 41 of the series „Monografías de Matemática".

Mathematics Subject Classification (1980): Primary: 58G25, 35P15, 52A40
Secondary: 58G11, 58C40, 58G30

ISBN 3-540-16788-9 Springer-Verlag Berlin Heidelberg New York
ISBN 0-387-16788-9 Springer-Verlag New York Berlin Heidelberg

Library of Congress Cataloging-in-Publication Data. Bérard, Pierre H. Spectral geometry. (Lecture notes in mathematics; 1207) Bibliography: p. Includes index. 1. Geometry, Riemannian. 2. Eigenvalues. 3. Operator theory. I. Title. II. Series: Lecture notes in mathematics (Springer-Verlag); 1207.
QA3.L28 no. 1207 [QA649] 510 s [516.3'73] 86-20323
ISBN 0-387-16788-9 (U.S.)


Printed in Germany

Printing and binding: Druckhaus Beltz, Hemsbach/Bergstr.
2146/3140-543210

To Rachel, Philippe, Izabel

# INTRODUCTION

The purpose of these notes is to describe some aspects of direct problems in spectral geometry.

Eigenvalue problems were motivated by questions in mathematical physics. In these notes, we deal with eigenvalue problems for the Laplace-Beltrami operator on a compact Riemannian manifold. To such a manifold $(M,g)$, we can associate a sequence of non-negative real numbers $\{\lambda_i\}_{i\geq 1}$, the eigenvalues of the Laplace-Beltrami operator $\Delta^g$ acting on $C^\infty(M)$. One can think of a Riemannian manifold as a musical instrument together with the musician who plays it. In this picture, the eigenvalues of the Laplace operator correspond to the harmonics of the instrument; they may depend on the music player, i.e. on the Riemannian metric: think of a kettledrum, or better of a Brazilian "cuíca".

Spectral geometry aims at describing the relationships between the musical instrument and the sounds it is capable of sending out.

The problems which arise in spectral geometry are of two kinds: direct problems and inverse problems. In a direct problem, we want information on the sounds produced by the instrument, in terms of its geometry. For example, we know that the bigger the tension of the parchment head of a kettledrum, the higher the pitch. In an inverse problem, we investigate what geometric information on the instrument can be recovered from the sounds it sends out.

Both types of problems are relevant to deep questions arising in mathematical physics (for example in elasticity theory, in plasma physics, in spectroscopy...).

This book could be divided into three parts: Chapters I to III; Chapters IV to VI and Appendix A; Chapter VII and Appendices B and C.

In Chapter I, we give some very simple-minded motivations from mathematical physics. Our purpose is not to derive mathematical models for some physical phenomena, but rather to show how some mathematical objects which will be introduced later on, arise naturally from physical principles. For further reading, we suggest [C-H] and [TL].

Chapter II is devoted to Riemannian geometry. We introduce the basic notions (geodesics, curvature,...) and we state, mainly without proofs, the basic results. In order to understand Chapter VI, the reader must have in mind the comparison theorems which involve the curvature of a Riemannian manifold. For further reading, we suggest [B-C], [CO], [C-E], [M-S] and [SI].

In Chapter III, we introduce the Laplace-Beltrami operator, and we describe the eigenvalue problems we will deal with in this book. An important part of this chapter is devoted to the variational characterizations of the eigenvalues. This is very important for later purposes. Although this material can be considered as classical ([KO], [R-S] or [C-H]), we have tried to describe it at length. The last paragraph of Chapter III contains general considerations on direct and inverse problems, and some answers to such problems as an illustration of the variational characterizations of the eigenvalues.

Chapters IV to VI form the core of this book. They contain results related to isoperimetric inequalities and to an important topic in Riemannian geometry, namely the interactions between local geometry (curvature estimates) and global geometry (topology...).

Many of the results we present in these chapters are new and are not yet available in print. These results were obtained in collaboration with G. Besson and S. Gallot (see [B-B-G1 to 3], [B-G]).

In Chapter IV, we introduce isoperimetric methods on compact Riemannian manifolds without boundary. The general setup described in § B, as well as the proof of J. Cheeger's lower bound for the first non-zero eigenvalue of a closed Riemannian manifold, are new. They arose from the above mentioned papers, and from brainstorming sessions with G. Besson and S. Gallot.

In Chapter V, we introduce the heat equation and then go directly to the main tool in this book: the isoperimetric inequality for the heat kernel. The ideas we develop here are those of [B-G]; our presentation differs however from that of [B-G] and is more in the spirit of Chapter IV.

Chapter VI is devoted to some applications of isoperimetric inequalities to Riemannian geometry. We use the ideas of [B-G], and the isoperimetric inequality obtained in [B-B-G1], to give bounds on topological invariants. The underlying method is the analytic method introduced by S. Bochner in the early 1940's, to obtain vanishing theorems. This method was improved by P. Li (1980) to give estimating theorems for Betti numbers, and later by S. Gallot (1981) to give estimating theorems in a more general framework. Both used isoperimetric estimates for Sobolev constants. In Chapter VI, we introduce a new idea (that of using Kato's inequality on heat kernels) which is due to M. Gromov, and came to life with the isoperimetric inequalities on the heat kernel given in [B-G]. It is important to read this chapter keeping in mind the compactness theorems of M. Gromov. These theorems are briefly described in the last paragraph of Chapter VI (see [SI] for a review).

These chapters are completed by an Appendix written by G. Besson.

In Appendix A, G. Besson shows how one can think of symmetrization procedures as relationships between Riemannian Geometry/Spectral Geometry on the one hand, and Operator Theory in a Hilbert space on the other hand; he views Kato's inequality (Chapter VI), and the symmetrization à la Faber-Krahn (Chapters IV and V), as particular cases of a unique general theorem. This interpretation is important because it distinguishes geometric techniques (isoperimetric inequalities) from analytic techniques (quadratic forms and operator theory); it also separates technicalities from fundamental ideas.

I am very grateful to G. Besson for writing this Appendix.

Spectral geometry has witnessed much research activity since the late 1960's. In Chapter VII, we very briefly sketch some of the important recent developments (in particular, see the very end of Chapter VII for more references).

Appendix B is a bibliography which I compiled in collaboration with M. Berger. I would like to thank M. Berger for allowing me to include it here. This bibliography is reproduced from the printed original; I thank the publisher Kaigai Publications (Japan) who left us the copyright. This bibliography is referred to as [B-B] in the text. It is divided into several chapters dealing with the different aspects of spectral geometry. Although the title refers to 1982, we revised the bibliography in September 1983. In Appendix C, I have added some new references.

This book was originally written both as a support for, and as a complement to lectures delivered at the 15º Colóquio Brasileiro de Matemática, July 1985. Although I have tried to give many complete proofs, I deliberately put emphasis on ideas rather than on technicalities. In a sense this book is an invitation to spectral geometry, rather than a course on spectral geometry. The original notes were first published by IMPA, in the series "Colóquio Brasileiro de Matemática". This new edition differs very little from the original one, as far as the mathematics are concerned: in order to avoid delay, I have only corrected some mistakes in the original text. In an attempt to make these notes more useful, I have added Appendix C (as a complement to [B-B]) and two indexes.

I thank the organizing Committee of the 15º Colóquio Brasileiro de Matemática for the opportunity to give a course on spectral geometry, and IMPA for its hospitality.

It is a pleasure for me to thank M.F. Cordel and P. Strazzanti who typed the first version of these notes, as well as Rogério Dias Trindade who typed the present text, for their care and competence.

I profited very much from regular brainstorming sessions with G. Besson and S. Gallot over the last three years. This book is an outgrowth of our collaboration. I owe them very much.

This book is dedicated to Marcel Berger in acknowledgement of his teachings.

Rio de Janeiro, April 1986.

# CONTENTS [1]



(1) Numbers refer to pages

## CHAPTER V

## ISOPERIMETRIC METHODS AND THE HEAT EQUATION



## CHAPTER VI

## GEOMETRIC APPLICATIONS OF ISOPERIMETRIC METHODS



## CHAPTER VII

## A BRIEF SURVEY OF SOME RECENT DEVELOPMENTS IN SPECTRAL GEOMETRY



## APPENDIX A by G. BESSON

## ON SYMMETRIZATION

# CHAPTER I

## MOTIVATIONS AND THE PHYSICAL POINT OF VIEW

1. The purpose of this chapter is to introduce some basic concepts which arise naturally from problems in mathematical physics. Our presentation might appear childish...; we do not aim at establishing good mathematical models for some elasticity problems. We only want to show how the notions of energy integral, variational methods, boundary conditions, wave equation, separation of variables, eigenvalue problems... arise naturally from problems in mathematical physics, and how they are related to other fields in mathematics (partial differential equations, spectral theory, Riemannian geometry).

### A. AN ELEMENTARY EXAMPLE

2. Let us consider a homogeneous elastic string $S$ whose position at rest is represented by the line segment $[0,L]$ in the plane. The string being elastic, the tension forces are tangential to the string. The string being homogeneous, the linear density $\rho$ and the tension $\mu$ of the string are constant along the string.

The first problem we shall deal with is that of the equilibrium position of the string $S$ submitted to an external force which acts in the plane, transversally to the string, with intensity $f(x)$.

We represent the equilibrium position of the string by a function $u: [0,L] \to \mathbb{R}$, the amplitude of the deflection of the string, therefore assuming that the points of the string can only move transversally.

The potential energy of the string consists of two terms: the energy $E_t(u)$ which arises from the tension $\mu$, and the external energy $E_e(u)$ which arises from the force applied to the string. The energy $E_t$ equals the tension times the increase of length of the string; the external energy is the work of the force $f$. We have

$$(3)\qquad \begin{cases} E_t(u) = \mu[\displaystyle\int_0^L (1+(u_x')^2)^{1/2}dx - L]; \\ \\ E_e(u) = \displaystyle\int_0^L f(x)u(x)dx. \end{cases}$$

We shall now make the assumption that the deflection of the string is "very small" in the sense that we can replace $(1+(u_x')^2)^{1/2}$ by $\frac{1}{2}(u_x')^2$. The potential energy of the string can then be replaced by

$$(4)\qquad E(u) = \frac{\mu}{2}\int_0^L (u_x')^2dx + \int_0^L f(x)u(x)dx.$$

In order to find $u$, we apply the <u>principle of least potential energy</u> which says that a stable equilibrium $u$ is a local minimum of the energy $E$, which implies that

$$(5)\qquad \frac{d}{d\varepsilon}E(u+\varepsilon v)\Big|_{\varepsilon=0} = 0,$$

where $u + \varepsilon v$ represents a position of the string close to the equilibrium $u$.

If we plug condition (5) into (4), we find

$$(6)\qquad \mu\int_0^L u_x'v_x'dx + \int_0^L f(x)v(x)dx = 0.$$

We can of course take local variations $v$, i.e. variations with compact support in $]0,L[$. Taking such a variation and integrating by parts, we find that for all $v$ in $C_0^\infty(]0,L[)$,

$$\int_0^L [-\mu u''_{xx} + f(x)]v(x)dx = 0 \text{ and hence}$$

(7) $$\mu \frac{d^2u}{dx^2}(x) = f(x) \quad \text{in} \quad ]0,L[.$$

8. <u>Remark</u>. We have implicitely made the assumption that $u$ is twice differentiable, in order to be able to write (7). We shall show how one can make weaker assumptions later on (nº 43).

9. Let us now take the function $v$ in $C^\infty([0,L])$. Equation (6) becomes, after integration by parts,

$$\mu u'_x v\Big|_0^L + \int_0^L (f(x)-\mu u''_{xx})\, v(x)dx = 0.$$

Taking (7) into account, we then have

(10) $$u'_x(L)v(L) - u'_x(0)v(0) = 0.$$

The fact that one can take one $v$ or another depends on the physical problem at hand. If we do not impose any condition on $v$, we deduce from (10) that $u$ must satisfy the <u>natural boundary condition</u> (Neumann boundary condition)

(10N) $$u'_x(0) = 0 \quad \text{and} \quad u'_x(L) = 0.$$

If we assume that the string is fixed at both ends (think of a violin or a piano string), we must impose that the deflection of the string is $0$ at $x = 0$ and $x = L$. This means that both $u$ and $v$ must satisfy the boundary condition (Dirichlet boundary condition)

(10D) $u(0) = 0$ and $u(L) = 0$.

In that case, (10) is void. The boundary condition (10N) corresponds to a <u>free string</u>, for which all deflections are allowed or <u>admissible</u>. The boundary condition (10D) corresponds to a string which is <u>fixed at both</u> ends. We then impose that the deflections satisfy $u(0) = 0$ and $u(L) = 0$. It is physically very intuitive that such conditions must be imposed to determine the equilibrium position of the problem under consideration.

11. <u>Summary</u>. In order to determine the equilibrium of a string submitted to a transversal external force $f$, we can

(i) either seek the local extrema of the energy

$$E(u) = \frac{\mu}{2}\int_0^L (u_x')^2 dx + \int_0^L f(x)u(x)dx,$$

when $u$ varies in a space of <u>admissible functions</u>, corresponding to the physical problem under consideration;

(ii) or solve the equation

$$\mu \frac{d^2u}{dx^2}(x) = f(x) \quad \text{in} \quad ]0,L[,$$

where some <u>boundary conditions</u> are imposed to $u$ at $x = 0$ and $x = L$, depending on the problem which is considered.

<u>Examples</u>:

<u>Dirichlet problem</u> (string fixed at both ends):

- Admissible functions: $u \in C^2([0,L])$ (see nº 8) such that $u(0) = u(L) = 0$ ($u+\epsilon v$ must also be admissible),
- Boundary conditions: $u(0) = 0$ and $u(L) = 0$;

<u>Neumann problem</u> (free string)

. Admissible functions: $u \in C^2([0,L])$ (see nº 8),

. Boundary conditions: $u'(0) = 0$ and $u'(L) = 0$
(imposed by the least potential energy principle).

12. Let us now consider the problem of the <u>vibrating string</u>, i.e. let us determine the laws of motion of an elastic string. We denote by $u: \mathbb{R} \times [0,L] \to \mathbb{R}$ the deflection of the string which is assumed to be transverse and small (in the sense used to derive (4)). The function $f$ considered above may also depend on the time parameter $t$. We then have to consider the kinetic energy of the string, namely

$$E_k(u) = \int_0^L \frac{1}{2}\,\rho(u_t')^2(t,x)dx. \tag{13}$$

Let $t_1$ and $t_2$ be two instants of time. <u>Hamilton's principle</u> states that the motion $u(t,x)$ of the string between the instants of time $t_1$ and $t_2$ should minimize the expression

$$J(u) = \int_{t_1}^{t_2}\int_0^L \{\frac{1}{2}\rho(\frac{\partial u}{\partial t}(t,x))^2 - \frac{1}{2}\,\mu(\frac{\partial u}{\partial x})^2(t,x) - f(t,x)u(t,x)\}dtdx,$$

among all <u>admissible motions</u> close to $u$, taking the same values as $u$ at $t = t_1$, and $t = t_2$ i.e.

$$\frac{d}{d\varepsilon}J(u+\varepsilon v)\Big|_{\varepsilon=0} = 0, \tag{14}$$

for all admissible functions $v$ such that $v(t_1,x) = 0$ and $v(t_2,x) = 0$, for all $x$ in $[0,L]$.

The adjective <u>admissible</u> refers to functions describing the physical problem under consideration as above (see nº 9 to 11).

Applying Hamilton's principle with $v \in C^\infty(\mathbb{R} \times [0,L])$ satisfying $v(t_1,x) = 0$, $v(t_2,x) = 0$, for all $x$, and integrating by parts, we deduce from (14) that

$$\int_{t_1}^{t_2} \int_0^L \{\rho \frac{\partial^2 u}{\partial t^2}(t,x) - \mu \frac{\partial^2 u}{\partial x^2}(t,x) + f(t,x)\} \, v(t,x) dt \, dx$$

$$+ \int_{t_1}^{t_2} \mu \frac{\partial u}{\partial x}(t,x) v(t,x) dt \Big|_0^L = 0.$$

The choice of $v$ being arbitrary we conclude that

$$\rho \frac{\partial^2 u}{\partial t^2}(t,x) - \mu \frac{\partial^2 u}{\partial x^2}(t,x) + f(t,x) = 0 \quad \text{in} \quad \mathbb{R} \times ]0,L[, \tag{15}$$

$$\frac{\partial u}{\partial x}(t,x) v(t,x) \Big|_0^L = 0 \quad \text{for all admissible} \quad v, \quad \text{and all} \quad t. \tag{16}$$

In the case of a string with free ends (i.e. no condition on $u$ and $v$), Equation (16) gives (Neumann conditions)

$$\frac{\partial u}{\partial x}(t,0) = 0 \quad \text{and} \quad \frac{\partial u}{\partial x}(t,L) = 0 \quad \text{for all} \quad t. \tag{16N}$$

In the case of a string with fixed ends, we must impose $u(t,0) = u(t,L) = 0$ and $v(t,0) = v(t,L) = 0$ for all $t$. Equation (16) is then always satisfied, and we only write the condition that $u$ is admissible (Dirichlet conditions)

$$u(t,0) = 0 \quad \text{and} \quad u(t,L) = 0 \quad \text{for all} \quad t. \tag{16D}$$

Equation (15) is called the one-dimensional <u>wave equation</u> (the space variable $x$ being one-dimensional).

17. <u>Remark</u>. In order to be able to determine the motion $u(t,x)$ of the string, we need Equation (15), boundary conditions e.g. (16D) or (16N) and <u>initial conditions</u>; these initial conditions already appeared in the statement of Hamilton's principle; we also consider the <u>Cauchy data</u> $u(t_o,x) = u_o(x)$ and $u_t'(t_o,x) = u_1(x)$, $0 \le x \le L$,

which describe the string at time $t_o$.

18. Summary. In order to determine the motion of a vibrating string submitted to a transversal external force $f$, we can

(i) either seek the extrema of the integral

$$J(u) = \int_{t_1}^{t_2}\int_0^L \{\frac{1}{2}\rho(\frac{\partial u}{\partial t})^2 - \frac{1}{2}\mu(\frac{\partial u}{\partial x})^2 - fu\}dtdx,$$

when $u$ varies in a space of admissible functions corresponding to the physical problem under consideration;

(ii) or solve the equation

$$\rho\frac{\partial^2 u}{\partial t^2} - \mu\frac{\partial^2 u}{\partial x^2} - f = 0 \quad \text{in} \quad \mathbb{R}\times\, ]0,L[,$$

$$\text{with initial conditions} \quad \begin{cases} u(t_o,x)=u_o(x) \\ \\ u_t'(t_o,x)=u_1(x) \end{cases}, \quad x \text{ in } [0,L],$$

and boundary conditions at $x = 0$ and $x = L$. (e.g. Dirichlet or Neumann conditions described in nº 11).

We can reduce the problem of the equilibrium to the present one by making all functions independent of the time variable $t$.

19. Some Comments

(a) Equations and boundary conditions: the transversal force acting on the string could also be related to the deflection $u(t,x)$ e.g. this could be an elastic force proportional to $u(t,x)$; we could also assume that there is a force acting on the ends of the string e.g. the ends could be elastically attached instead of fixed. Such conditions can give rise to other contributions to the energy of the string, thus modifying both the equation and the boundary conditions.

(b) Considering **elastic bars instead** of strings, we would arrive at the following situation

$$(20)\quad \begin{cases} \cdot\ J(u) = \displaystyle\int_{t_1}^{t_2}\int_0^L \{\frac{1}{2}\rho(x)(\frac{\partial u}{\partial t})^2 - \frac{1}{2}\mu(x)(\frac{\partial u}{\partial x})^2 - fu\}dt\,dx, \\ \cdot\ \rho(x)\dfrac{\partial^2 u}{\partial t^2} - \dfrac{\partial}{\partial x}(\mu(x)\dfrac{\partial u}{\partial x}) - f = 0, \end{cases}$$

where $\rho(x)$ is the linear density of the bar and where $\mu(x)$ describes the elasticity of the bar (both functions are positive). We assume these functions to depend on the space variable $x$, but not on the time variable.

<u>References</u>

[C-H] Chap. IV § 10, p. 242 ff,

[FO] Chap. 5 p. 130 ff (p. 168 for the elastic bar),

[WR] Chap. I, for both the elastic string and the elastic bar.

## B. THE METHOD OF SEPARATION OF VARIABLES

21. In order to study the wave equation which appears in (20), it is convenient to look at a simpler problem, namely the case $f \equiv 0$, and to seek solutions $u(t,x)$ of the form $F(t)G(x)$ (i.e. to <u>separate variables</u>, a method which goes back to the 18th century). Equation (20) becomes

$$(22)\qquad \rho(x)G(x)\frac{d^2F}{dt^2}(t) - F(t)\frac{d}{dx}(\mu(x)\frac{dG}{dx}(x)) = 0,$$

which is easily seen to split into two equations

$$(23)\quad \begin{cases} \text{(i)} & \dfrac{d}{dx}(\mu(x)\dfrac{dG}{dx}(x)) + \lambda\rho(x)G(x) = 0, \quad x \in ]0,L[, \\ \\ \text{(ii)} & \dfrac{d^2F}{dt^2}(t) + \lambda F(t) = 0, \quad t \in \mathbb{R}, \end{cases}$$

for some constant $\lambda$.

If we now recall that $u(t,x)$ must be an admissible function, e.g. that it satisfies one of the boundary conditions (16D) or (16N), we have to impose boundary conditions on $G$, e.g.

$$(24)\quad \begin{cases} G(0) = G(L) = 0 & \text{(Dirichlet conditions),} \\ \text{or} & \\ G'(0) = G'(L) = 0 & \text{(Neumann conditions).} \end{cases}$$

Let us for example consider the Dirichlet boundary conditions. We are led to the Sturm-Liouville problem

$$(25)\quad \begin{cases} (\mu(x)G'(x))' + \lambda\rho(x)G(x) = 0, \\ \\ G(0) = G(L) = 0. \end{cases}$$

It can be shown ([SR] Chap. IV or [C-H]) that the $\lambda$'s for which (25) has a non-trivial solution form an infinite sequence $\lambda_1 < \lambda_2 < \ldots \uparrow +\infty$ of positive real numbers going to infinity (these numbers are called the eigenvalues of the Sturm-Liouville Problem (25)). To the eigenvalue $\lambda_n$ of Problem (25) corresponds a one-dimensional space of eigenfunctions.

We can choose an eigenfunction $G_n$ corresponding to $\lambda_n$, normalized by $\int_0^L \rho(x)G_n^2(x)dx = 1$. The basic fact is that a given function $f(x)$ can, under certain mild conditions, be represented by an infinite series in the $G_n$'s; $f(x) = \sum_{n=1}^{\infty} a_n G_n(x)$.

The case of Fourier-sine series is a particular instance of this fact ($\mu=1$, $\rho=1$). Let us make some formal computations. The functions which appear in Equation (20) can be written as infinite series in the $G_n$'s (as far as the $x$ variable is concerned); we thus have (summations from 1 to $\infty$)

$$f(t,x) = \Sigma\, a_n(t)G_n(x);$$
$$u(t,x) = \Sigma\, b_n(t)G_n(x);$$
$$u(t_o,x) = u_o(x) = \Sigma\, c_nG_n(x);$$
$$u'_t(t_o,x) = u_1(x) = \Sigma\, d_nG_n(x).$$

At least at the formal level, plugging these series into Equation (20), we obtain

$$(26)\qquad \left\{\begin{array}{l} b''_n(t) + \lambda_n b_n(t) = a_n(t) \\ b_n(t_o) = c_n \\ b'_n(t_o) = d_n \end{array}\right\}, \quad n \text{ in } \mathbb{N}.$$

Since it is easy to solve (26), we have an expression of $u(t,x)$ in terms of series representing $f(t,x)$, $u(t_o,x)$ and $u'_t(t_o,x)$.

These formal calculations explain why it is so important to determine the eigenvalues of the Sturm-Liouville Problem (25). In these notes, we shall deal with generalizations of the situation we have just described.

For more details on Sturm-Liouville problems and their eigenfunctions expansions, we refer to [C-H], [SR] Chap. IV and [FO] (for the case of Fourier series), or [D-M].

## C. GENERALIZATIONS

27. Let us now consider a vibrating homogeneous membrane, whose position at rest is represented by a bounded, regular domain $\Omega$ in $\mathbb{R}^2$.

We are again interested in transverse vibrations of the membrane (i.e. normal to the plane $\mathbb{R}^2$). We denote by $u(t,x)$, $(t,x) \in \mathbb{R} \times \Omega$, the amplitude of such a vibration.

In order to make things simpler, we shall assume that no external force acts on the membrane, and that the membrane is either fixed on its boundary $\partial\Omega$ (this is a drum) or free. The corresponding <u>admissible functions</u> in the sense of nº 10-11 are in $C^2(\Omega)$, and assumed to vanish on the boundary $\partial\Omega$ when the membrane is fixed (no condition when the membrane is free). We denote the density of the membrane by $\rho$, and its tension by $\mu$. The kinetic energy of the membrane is given by

$$(28.1) \qquad E_k(u) = \frac{1}{2}\rho \int_\Omega \left(\frac{\partial u}{\partial t}\right)^2(t,x)dx,$$

and the potential energy is given by $\mu$ times the increase of area of the membrane, i.e.

$$\mu \int_\Omega [(1+|\nabla_x u|^2(t,x))^{1/2} - 1]\, dx,$$

which we shall approximate (under the assumption that the vibration is "small", compare with nº 3-4) by

$$(28.2) \qquad E_p(u) = \frac{1}{2}\mu \int_\Omega |\nabla u|^2(t,x)dx,$$

where $\nabla u$ is the gradient of $u$ in the x-variable, i.e. (in Cartesian coordinates)

$$\nabla u(t,x) = \left(\frac{\partial u}{\partial x_1}(t,x), \frac{\partial u}{\partial x_2}(t,x)\right), \quad \text{if} \quad x = (x_1,x_2),$$

and where $|x|^2 = x_1^2 + x_2^2$, for $x \in \mathbb{R}^2$.

In order to derive the laws of motion of the membrane, we again use Hamilton's principle.

We define

$$(29) \qquad J(u) = \frac{1}{2}\int_{t_1}^{t_2}\int_{\Omega}[\rho(\frac{\partial u}{\partial t})^2(t,x) - \mu|\nabla u|^2(t,x)]dt\,dx,$$

and we seek admissible functions $u$, such that for all admissible functions $v$, with $v(t_1,x) = 0$, $v(t_2,x) = 0$ for all $x$, we have

$$(30) \qquad \frac{d}{d\varepsilon}J(u+\varepsilon v)\Big|_{\varepsilon=0} = 0.$$

If we plug (29) into (30), we obtain

$$\int_{t_1}^{t_2}\int_{\Omega}[\rho\,\frac{\partial u}{\partial t}\,\frac{\partial v}{\partial t} - \mu(\nabla u|\nabla v)]dt\,dx = 0,$$

where $(.|.)$ is the scalar product in $\mathbb{R}^2$.

If we apply integration by parts in the t-variable, and Green's formula in the x-variable ($n$ being the inner unit normal to $\partial\Omega$), we obtain

$$(31) \qquad \int_{t_1}^{t_2}\int_{\Omega}(\rho\,\frac{\partial^2 u}{\partial t^2} + \mu\Delta u)v\,dt\,dx - \int_{t_1}^{t_2}\int_{\partial\Omega}\mu v(\nabla u|n)dt\,d\sigma = 0$$

where $\Delta u = -(\frac{\partial^2 u}{\partial x_1^2} + \frac{\partial^2 u}{\partial x_1^2})$ (<u>Note our sign convention</u>), and

$d\sigma$ = arc length on $\partial\Omega$.

If we take $v$ with compact support in the x-variable (inside $\Omega$), we deduce that $u$ must satisfy the two-dimensional wave equation

$$(32) \qquad \rho\,\frac{\partial^2 u}{\partial t^2}(t,x) + \mu\Delta u(t,x) = 0 \quad \text{in } \mathbb{R}\times\Omega.$$

If we deal with a fixed membrane, we have to impose the condition that $u$ and $v$ vanish on $\partial\Omega$, so that the second term in (31) is always $0$; we then have the <u>boundary condition</u> (Dirichlet condition)

(33D) $\quad u(t,x) = 0$ for all $(t,x) \in \mathbb{R} \times \partial\Omega$.

If we deal with a free membrane, we can take a $v$ in $C^{\infty}(\mathbb{R}\times\Omega)$ so that (31) and (32) imply that $u$ must satisfy the <u>natural boundary condition</u> (Neumann condition)

(33N) $\quad (\nabla u|n) = 0$ on $\mathbb{R} \times \partial\Omega$

(We shall write $\frac{\partial u}{\partial n} = (\nabla u|n)$).

For more details, we refer the reader to [C-H] Chap. IV §10 p. 242 ff and [PY] p.7.

As for the vibrating string, we can now apply the method of separation of variables, and we have to deal with the following problem

$$(34) \quad \begin{cases} \Delta U(x) = \lambda U(x) & \text{in } \Omega, \\ U(x) = 0 & \text{on } \partial\Omega \quad (\text{for (33D)}), \\ \text{or} & \\ \frac{\partial U}{\partial n}(x) = 0 & \text{on } \partial\Omega \ (\text{for (33N)}). \end{cases}$$

Problem (34) is far more difficult than its one-dimensional analogue (25). As was shown by H. Poincaré at the end of the $19^{\underline{th}}$ century, Problem (34) admits a non-trivial solution for values of $\lambda$ which form an infinite sequence of non-negative numbers, which goes to infinity, $(0\le)\lambda_1 \le \lambda_2 \le \dots$. Given an <u>eigenvalue</u> $\lambda_n$ of (34), the vector space formed by the solutions of Equation (34) with $\lambda=\lambda_n$ is <u>finite dimensional</u> (its elements are called <u>eigenfunctions</u> associated with $\lambda_n$, and its dimension the <u>multiplicity</u> of $\lambda_n$).

In these notes we shall be interested in problems similar to Problem (34) with $\Omega$ (a domain in) a differentiable manifold, and

$\Delta$ an operator which will generalize the ordinary Laplacian in $\mathbb{R}^2$.

We shall not go into any further details now. The reader interested in Problem (34) may read the appropriate chapters of [TS].

35. Let us now indicate a generalization of the above situation. In certain problems of elasticity, dealing with non-homogeneous media, one has to consider an expression of the potential energy of the following form

$$E_p(u) = \int_\Omega Q(x, \nabla u)\, dx, \tag{36}$$

where $Q(x,.)$ is a positive definite quadratic form on $\mathbb{R}^2$ whose coefficients are functions of the space variable (in some mechanical problems $Q(x,.)$ describes the tensor of constraints). If we plug the expression (36) into (29), and if we apply Hamilton's principle, we obtain an equation similar to (32) with an operator $\Delta_Q$ which generalizes $\Delta$. We shall meet such expressions later, $Q(x,\nabla u)$ will then be associated with some <u>Riemannian metric</u> on the manifold $\Omega$.

D. OTHER POINTS OF VIEW

Let us now look back at what we did in paragraphs A to C. The energy or Dirichlet integral

$$(37)\quad \begin{cases} \text{(i)}\quad E(u) = \int_0^L (u'_x)^2 dx & \text{(in the case of the vibrating string),} \\ \text{or} & \\ \text{(ii)}\quad E(u) = \int_\Omega |\nabla u|^2 dx & \text{(in the case of the vibrating membrane),} \end{cases}$$

plays a prominent role. To this energy integral, the <u>variational approach</u> associates the Laplacian

$$(38)\quad \begin{cases} \text{(i)}\quad \Delta = -\dfrac{d^2}{dx^2} \quad \text{or} \quad -\dfrac{d}{dx}(\mu(x)\dfrac{d}{dx}) & \text{(one-dimensional case),} \\[2ex] \text{(ii)}\quad \Delta = -(\dfrac{\partial^2}{\partial x_1^2} + \dfrac{\partial^2}{\partial x_2^2}) & \text{(two-dimensional case).} \end{cases}$$

(We use the minus sign for convenience; we shall keep this convention throughout this text).

Having in mind the eigenvalue Problems (25) and (34), and recalling Lagrange's multipliers method, we also introduce the Rayleigh (-Ritz) quotient

$$(39)\quad \begin{cases} \text{(i)}\quad R(u) = \displaystyle\int_0^L (u_x')^2 dx \Big/ \int_0^L u^2 dx & \text{(one-dimensional case),} \\[2ex] \text{(ii)}\quad R(u) = \displaystyle\int_\Omega |\nabla u|^2 dx \Big/ \int_\Omega u^2 dx & \text{(two-dimensional case),} \end{cases}$$

where $u$ is not identically zero. Indeed, if we write

$\frac{d}{d\epsilon} R(u+\epsilon v)\Big|_{\epsilon=0} = 0$, we find, say with (39ii),

$$(40)\quad \int_\Omega \langle \nabla u, \nabla v\rangle \, dx = R(u) \int_\Omega uv \, dx.$$

Assume that (40) holds for all functions $v$ in $C_0^\infty(\Omega)$, and let $R(u) = \lambda$. Integrating by parts gives

$$\Delta u = \lambda u \quad \text{in} \quad \Omega \quad (\Delta \text{ as in (38ii)!}).$$

SUMMARY - THE MAIN CHARACTERS OF THIS PLAY ARE

(41)

- the energy or Dirichlet integral $\int_\Omega |\nabla u|^2 dx$,
- the (positive) Laplacian $\Delta$ (positive refers to the sign convention made above nº (38) and (47ii) below),
- the Rayleigh (-Ritz) quotient $R(u) = \int_\Omega |\nabla u|^2 dx \,/ \int_\Omega u^2 dx$.

42. In the preceding paragraphs, we have shown how the partial differential equations governing the vibrations of an elastic string or membrane can be deduced from Hamilton's principle, once we know the expression of the energy. These partial differential equations involve the Laplacian $\Delta$. The method of separation of variables led us to some eigenvalue problems for the Laplacian. These eigenvalue problems are related to the extrema of the Rayleigh quotient $R(u)$ (or equivalently to the extrema of the energy $\int_\Omega |\nabla u|^2 dx$, under the constraint $\int_\Omega u^2 dx = 1$).

We shall now explain how these considerations are related to other points of view or formulations.

43. Let us first deal with the <u>point of view of partial differential equations</u> (P.D.E). Let $\Omega$ denote an elastic membrane which is fixed along $\partial\Omega$ and submitted to a transversal force $f$. In order to find the equilibrium position of the membrane, we have to look for local extrema of the energy

$$E(u) = \int_\Omega |\nabla u|^2 dx + \int_\Omega f(x)u(x)dx.$$

This leads us to the boundary value problem (compare with §A; the admissible functions are required to vanish on $\partial\Omega$)

$$(44) \qquad \begin{cases} \Delta u(x) + f(x) = 0 \quad \text{in } \Omega, \text{ with the boundary condition} \\ u(x) = 0 \quad \text{on } \partial\Omega. \end{cases}$$

Multiplying (44) by any function $v$ in $C_o^\infty(\Omega)$ we obtain, after integration by parts,

$$
(45)\quad \begin{cases} \text{(i)} \quad \int_\Omega \langle \nabla u, \nabla v \rangle dx + \int_\Omega f(x)v(x)dx = 0 \\ \text{(ii)} \quad \int_\Omega u\Delta v \, dx + \int_\Omega f(x)v(x)dx = 0 \end{cases}
$$

If $u$ is twice differentiable and satisfies (44), we say that $u$ is a <u>classical solution</u> of Equation (44). If $u$ is in $L^1_{loc}(\Omega)$ and satisfies Equation (45ii), we say that $u$ is a solution of Equation (44) in the <u>sense of distributions</u>. If $u$ and $|\nabla u|$ are in $L^2(\Omega)$ and if $u$ satisfies (45i), we say that $u$ is a <u>weak solution</u> of Equation (44). We do not want to go into technical details here, for precise definitions and results see [TS], [G-T] or [SW].

It turns out that it is much easier to prove the existence of a weak solution than that of a classical solution. Once the existence of a weak solution is proved (e.g. by using Hilbert space methods and appropriate Sobolev spaces), one has to prove that the weak solution is indeed a classical solution: one has to prove interior regularity in $\Omega$, and regularity up to $\partial\Omega$. Note that the bilinear form $\int_\Omega \langle \nabla u, \nabla v \rangle dx$ in Equation (45i) is just the bilinear form associated with the quadratic form giving the energy, $\int_\Omega |\nabla u|^2 dx$.

46. We have seen (nº 21-26) that it is very important to solve the eigenvalue problem $\Delta u = \lambda u$ in $\Omega$, with some appropriate boundary condition, e.g. the Dirichlet boundary condition $u = 0$ on $\partial\Omega$. The Laplacian $\Delta$ is a linear (partial differential) operator. We could view it as a linear operator from $C^2(\Omega)$ into $C^o(\Omega)$ but this is not so good if we want to consider the eigenvalues of $\Delta$. We

could also consider $\Delta$ as a linear operator from $C^\infty(\Omega)$ into $C^\infty(\Omega)$. It turns out that this is not an appropriate choice because the $C^\infty$-topology is too complicated and because Equation (45i) is so much related to the $L^2$-inner product in $\Omega$, $(u,v) \to \int_\Omega u(x)v(x)dx = (u|v)$ (if we deal valued functions or $(u,v) \to \int_\Omega u(x)\bar{v}(x)dx$, if we deal with complex valued functions). It turns out that the good choice is to view $\Delta$ as an unbounded linear operator on $L^2(\Omega)$, with <u>domain</u> $C_0^\infty(\Omega)$; this means that we consider $\Delta$ as a linear operator from the dense linear subspace $C_0^\infty(\Omega)$ of $L^2(\Omega)$ into $L^2(\Omega)$. Spectral theory was devised to deal with such operators. The Laplacian has the following properties

(47i) $\quad \forall\, u,v \in C_0^\infty(\Omega)\ (\Delta u|v) = (u|\Delta v)$,

we say that $\Delta$ is a <u>symmetric operator</u>;

(47ii) $\quad \forall\, u \in C_0^\infty(\Omega)\ (\Delta u|u) = \int_\Omega |\nabla u|^2 \geq 0$,

we say that $\Delta$ is a <u>positive operator</u>.

It follows from a theorem of Friedrichs ([R-S] Vol. II) that $\Delta$ can be extended to an <u>unbounded self-adjoint</u> operator $\Delta_e$ in $L^2(\Omega)$. The manner in which this extension is made depends on the boundary conditions which are imposed on $\partial\Omega$.

In the finite dimensional case, there is a very strong relationship between self-adjoint operators and quadratic forms. This can be generalized to more complicated situations. For example, in order to study the Laplacian $\Delta$ one can study the quadratic form given by the energy integral $u \to \int_\Omega |\nabla u|^2 dx$ (see Formula 47ii).

Let us deal with an example. If we want to study the eigenvalue problem

$$(48)\qquad \begin{cases} \Delta u = \lambda u & \text{in } \Omega, \\ u(x) = 0 & \text{on } \partial\Omega, \end{cases}$$

we look at the Rayleigh quotient $R(u) = \dfrac{\int_\Omega |\nabla u|^2 dx}{\int_\Omega u^2\, dx}$.

In order to study $R(u)$, we have to determine what are the admissible functions. For $u$ in $C^\infty(\Omega)$, we define $\|u\|_1^2 = \int_\Omega u^2(x)dx + \int_\Omega |\nabla u(x)|^2 dx$. We call $H^1(\Omega)$ (resp. $H_o^1(\Omega)$) the completion of $C^\infty(\Omega)$ (resp. $C_o^\infty(\Omega)$) for the norm $\|.\|_1$. Let $\|.\|_o$, $\|u\|_o^2 = \int_\Omega u^2(x)dx$, be the $L^2$-norm. Since $\|u\|_o \le \|u\|_1$ for $u \in C^\infty(\Omega)$, we have

$$H_o^1(\Omega) \subset H^1(\Omega) \subset L^2(\Omega).$$

The admissible functions for the Dirichlet problem are the functions in $H_o^1(\Omega)$ (we would take $H^1(\Omega)$ for the Neumann problem). Since $\Omega$ is a compact set, the inclusion $H_o^1(\Omega) \subset L^2(\Omega)$ is compact (this is the case for $H^1(\Omega) \subset L^2(\Omega)$, under some regularity conditions on $\partial\Omega$).

From this it follows that

$$\lambda_1(\Omega) = \inf\{R(u) \mid u \neq 0,\ u \in H_o^1(\Omega)\}$$

is achieved on a subspace $E_1(\Omega)$ of $H_o^1(\Omega)$; $E_1(\Omega)$ is characterized by the property

$$u \in E_1(\Omega) \Leftrightarrow \forall\ v \in H_o^1(\Omega),\ \int_\Omega \langle \nabla u, \nabla u\rangle dx = \lambda_1(\Omega) \int_\Omega uv\ dx.$$

In order to find the other eigenvalues, one has to consider

the orthogonal $H^{(1)}$ of $E_1(\Omega)$ in $L^2(\Omega)$. One then defines $\lambda_2(\Omega) = \inf\{R(u) \mid u \neq 0\ u \in H_o^1(\Omega) \cap H^{(1)}\}$... See nº III. 18 ff.

This manner of dealing with the eigenvalue problem (48) is very close to the underlying physical properties.

For example, consider two membranes $\Omega_1, \Omega_2$ with the same physical properties and such that $\Omega_1 \subset \Omega_2$. We then have $H_o^1(\Omega_1) \subset \subset H_o^1(\Omega_2)$, and we conclude that $\lambda_1(\Omega_1) \geq \lambda_1(\Omega_2)$: the smaller drum has a higher fundamental tone.

As was already alluded to before, a Riemannian metric $g$ on $\Omega$ may account for some physical properties (stress,...). Assume that one is given $\Omega$ with two Riemannian metrics $g_1$ and $g_2$ such that for any tangent vector $U$, $g_1(U,U) \leq g_2(U,U)$. Then, for any $u$ in $C_o^\infty(\Omega)$, one has $R(u;g_1) \geq R(u;g_2)$ (recall that $R$ involves the dual metric) and hence $\lambda_1(\Omega,g_1) \geq \lambda_1(\Omega,g_2)$.

We shall see in Chapter III nº 26 that there are variational characterizations of eigenvalues which are very similar to the one above for $\lambda_1(\Omega)$ (it is good to keep the finite dimensional case in mind). These characterizations are very important because they are very close to the original physical problems through the Rayleigh quotient.

It will be important in the sequel to keep in mind the physical motivations we described in this chapter.

Further references for Chapter I: Partial differential equations: [B-J-S],[ES],[GN],[GT],[PY],[TL],[TS],[WR],[WS]; Spectral theory: [AN],[R-S],[SW] and [BZ] for functional analysis; other possible references for motivations and results: [BE],[C-H],[CL],[TL].

CHAPTER II

# TOPICS FROM RIEMANNIAN GEOMETRY

## A. GENERALITIES

The purpose of this chapter is to introduce the basic objects we shall deal with in the core of these notes (Chapters IV to VI): Riemannian manifolds, curvatures, the covariant derivative...

The reader interested in Riemannian geometry itself is referred to [B-G-M] Chapters I and II, [C-E] or [CO], for more details and proofs.

All manifolds we shall consider will be $C^\infty$ connected manifolds (unless otherwise stated).

1. A Riemannian manifold $(M,g)$ is a manifold $M$ equipped with a Riemannian metric $g$: for any point $x$ in $M, g_x$ is a scalar product on the tangent space $T_xM$ which depends $C^\infty$ on $x$ (this can be checked in a local coordinate system).

2. Examples

(a) $(\mathbb{R}^n$, can): the space $\mathbb{R}^n$ equipped with the usual Euclidean structure is a Riemannian manifold; we can also consider $(\mathbb{R}^n, g_A)$, where $A$ is a $C^\infty$ map from $\mathbb{R}^n$ to the space $S_+(n)$ of positive definite symmetric $n \times n$ matrices on $\mathbb{R}^n$ and $g_A(x,y) = (Ax|y)$ for any vectors $x,y$ in $\mathbb{R}^n$ (here $(.|.)$ denotes the usual Euclidean structure). We can also restrict our attention to a smooth bounded domain $D$ in $\mathbb{R}^n$; in that case $g_A$ could represent a strain tensor inside the body D.

We call $(H^n, \text{can})$ the Riemannian manifold $(B, g_H)$ where $B$ is the open ball centered at 0, with radius 2 in $\mathbb{R}^n$; for $U, V$ tangent to $B$ at $x$, $g_H$ is defined by

$$g_H(U,V) = \left(1 - \frac{|x|^2}{4}\right)^{-2} (U|V)$$

where $|x|$ is the Euclidean norm of the vector $x$ in $\mathbb{R}^n$.

This Riemannian manifold is called the n-dimensional <u>hyperbolic space</u>.

(b) Let $f: M \to \mathbb{R}^N$ be an imbedding of a manifold $M$ into $\mathbb{R}^N$. The <u>induced metric</u> $g$ on $M$ is defined as the pull-back by $f$ of the canonical metric on $\mathbb{R}^N$; for any vectors $U$ and $V$ in $T_xM$, we define $g_x(U,V) = (f_*U|f_*V)$ the scalar product in $\mathbb{R}^N$ of the images of $U$ and $V$ by the tangent map $f_*$ to $f$. A very important instance of such a Riemannian submanifold of $\mathbb{R}^N$ is the <u>canonical sphere</u> $(S^n, \text{can})$, where $S^n$ is the unit sphere in $\mathbb{R}^{n+1}$ with induced Riemannian metric, $S^n = \{x \in \mathbb{R}^{n+1} \mid (x|x) = 1\}$ (For example $S^1$ in $\mathbb{R}^2$).

(c) The <u>Riemannian product</u> $(M \times N, g \times h)$ of two Riemannian manifolds $(M,g)$ and $(N,h)$ is defined in such a way that Phythagoras theorem be true: if $(U,V)$ (resp. $(U',V')$) are tangent vectors at $(x,y)$ in $M \times N$, then

$$(g \times h)((U,V),(U',V')) = g(U,U') + h(V,V').$$

For example, the n-torus $(T^n, \text{can})$ is the product of $n$ copies of $(S^1, \text{can})$ (see Example (b) and Example 4(d)).

3. <u>An isometry</u> $f: (M,g) \to (N,h)$ between two Riemannian manifolds is a diffeomorphism $f$ between $M$ and $N$ such that $f^*h = g$, i.e. for any $x$ in $M$ and $U$ in $T_xM$, $h_{f(x)}(f_*U, f_*U) =$ $= g_x(U,U)$.

4. Examples (continued)

(d) Let $(M,g)$ be a Riemannian manifold and let $G$ be a discrete group of isometries of $(M,g)$, such that the quotient space $M/G = N$ is a manifold.

It is then clear that one can define a Riemannian manifold $(N,h) = (M/G,g/G)$. For example $(T^n, \mathrm{can})$ defined in 2(c) is isometric to $(\mathbb{R}^n/\mathbb{Z}^n,\mathrm{can}/\mathbb{Z}^n)$. Other tori can be defined as follows: let $G$ be a lattice in $\mathbb{R}^n$ i.e. $G = e_1\mathbb{Z}+\ldots+e_n\mathbb{Z}$, where $[e_1,\ldots,e_n]$ is a basis of $\mathbb{R}^n$. We can define the torus $(T^n_G,\mathrm{can})$ as $(\mathbb{R}^n/G,\mathrm{can}/G)$. The tori $(T^n,\mathrm{can})$ and $(T^n_G,\mathrm{can})$ are not necessarily isometric (they are howewer always diffeomorphic).

Another instance of such a situation is the canonical Riemannian metric on the projective space $\mathbb{R}P^n$. We can view $\mathbb{R}P^n$ as the quotient of the sphere $S^n$ by the antipodal map which sends the point $x$ in $S^n$ to $-x$. We can then write $\mathbb{R}P^n = S^n/\{1,\sigma\}$. We define the Riemannian manifold $(\mathbb{R}P^n, \mathrm{can})$ as $(S^n/\{1,\sigma\},\mathrm{can}/\{1,\sigma\})$ because $\sigma$ is an isometry of $(S^n,\mathrm{can})$ (it is induced by the symmetry about $0$ in $\mathbb{R}^{n+1}$).

From the definition of the quotient Riemannian manifold $(M/G,g/G)$ it follows that one can also define a natural Riemannian metric on a covering space $M$ over a Riemannian manifold $(N,h)$.

For more details see [B-G-M] Chap. I, [CO] Chap. 1.

5. A Riemannian invariant is a function $F$ defined on the space of Riemannian metrics on a manifold $M$, which is invariant under isometries. This means that $F$ is in fact a function on the space of Riemannian structures on $M$, i.e. on the quotient space of the space of Riemannian metrics by the group of diffeomorphisms. For example, we do not necessarily want to view the Riemannian manifold $(S^n,\mathrm{can})$ as the unit sphere in $\mathbb{R}^{n+1}$ with induced metric; any other

isometric representation can serve. Any positive definite quadratic form on $\mathbb{R}^n$ with constant coefficients gives rise to the same Riemannian structure on $\mathbb{R}^n$ : $(\mathbb{R}^n, \text{can})$.

6. Scaling. Given a Riemannian manifold $(M,g)$, one has a whole family of Riemannian manifolds $(M,g_a)$ which are obtained from $(M,g)$ by multiplying the Riemannian metric $g$ by the positive constant $a$ : $g_a = ag$. A Riemannian invariant $F(g)$ may have a weight $r$, i.e. satisfy $F(ag) = a^r F(g)$. Since dilating the metric is very often a trivial operation, we shall be mainly interested in Riemannian invariants with weight 0. This will appear in a crucial way later on.

7. On a Riemannian manifold one can define the length of a curve $c: [0,1] \to M$ by

$$\ell(c) = \int_0^1 g(\dot{c}(t),\dot{c}(t))^{1/2} dt,$$

where $\dot{c}(t)$ is the velocity vector of the curve.

One can now define the Riemannian distance $d(x,y)$ or $\overline{xy}$ between two points $x$ and $y$ of $(M,g)$ as the infimum of the lengths of the curves in $M$ going from $x$ to $y$.

Caution. Let us consider $(S^2, \text{can})$ in $\mathbb{R}^3$. It shall be clear later on that the Riemannian distance between two antipodal points is $\pi$; we have to consider curves lying on $S^2$, not curves going through the ball. For this reason, the Riemannian distance on a submanifold of $\mathbb{R}^n$ is also referred to as the intrinsic distance (vs extrinsic distance).

8. Properties

(i) $d$ is a distance (in the sense of metric spaces) and

this distance defines the **same** topology on $M$ as the one given with the differentiable manifold structure;

(ii) A classical theorem of H. Hopf and W. Rinow ([CO] Chap. 7 or [B-C] § 8.2) states that if the metric space $(M,d)$ is complete, then any two points $x,y$ in $M$ can be joined by a curve (called <u>shortest path</u>) whose length is exactly $d(x,y)$.

9. <u>Variational arguments</u> show that a shortest path is carried by a <u>geodesic</u>. Geodesics are curves which satisfy a certain second order (non-linear) differential equation on $M$, see nº 41. Given $x$ in $M$ and $U$ in $T_xM$ there exists a unique geodesic $c_U$ starting from $x$ with velocity vector $U$ at $x$. An assertion in the Hopf-Rinow theorem states that $c_U(t)$ is defined for <u>all</u> values of $t$ if and only if $(M,d)$ is a complete metric space. In that case one says that $(M,g)$ <u>is a complete Riemannian manifold</u>.

From now on, <u>all Riemannian manifolds are assumed to be complete</u> (unless otherwise stated).

Geodesics are always parametrized proportionally to arc-length and are locally length minimizing (for $\epsilon$ small enough, $c$ is a shortest path between the points $c(t)$ and $c(t+\epsilon)$).

The geodesics of $(S^2,can)$ are the great circles. A shortest path between two points $x,y$ of $S^2$ is the piece of a great circle through $x$ and $y$ with smallest length. Antipodal points are joined by infinitely many shortest paths. Any two points of the sphere can be joined by at least two geodesics one of them of shortest length (two arcs of a great circle passing through $x$ and $y$).

10. The <u>diameter</u> $\mathrm{Diam}(M,g)$ of the Riemannian manifold $(M,g)$ is defined by

$$\mathrm{Diam}(M,g) = \sup\{d(x,y) : x,y \text{ in } M\}.$$

It is finite if and only if $M$ is compact. (As indicated above $(M,g)$ is already assumed to be complete).

11. Let $\{x_1,\ldots,x_n\}$ be local coordinates near a point $p$ in $M$. In these coordinates, the metric $g$ can be represented by the matrix $(g_{ij})$, $g_{ij} = g(\frac{\partial}{\partial x_i}, \frac{\partial}{\partial x_j})$. The measure $[\mathrm{Det}((g_{ij}))]^{1/2} dx_1 \ldots dx_n$ does not depend on the choice of a local coordinate system (use the theorem on change of variables in an integral). It defines the canonical Riemannian measure which we shall denote by $v_g$ (or sometimes simply by $dx$) ([B-G-M] Chap. II.A). Given a continuous function $f$ on $M$, we shall write $\int_M f(x)dv_g(x)$, $\int_M f dv_g$ or simply $\int_M f$ for the integral of the function $f$ on $M$.

12. Properties Let $(M,g)$ be a Riemannian manifold. Then

(i) $v_{ag} = a^{n/2} v_g$;

(ii) $\mathrm{Diam}(M,ag) = a^{1/2}\,\mathrm{Diam}(M,g)$

if $\dim M = n$ $(a > 0)$; see nº 6.

13. Given a point $x$ in a complete Riemannian manifold $(M,g)$, define the exponential map at the point $x$, $\exp_x : T_xM \to M$ as follows.

Given a vector $U$ in $T_xM$ we define $\exp_x(U)$ as the point $c_U(1)$ on the geodesic $c_U$ issued from $x$ with initial velocity vector $U$.

For nº 7-13, see [B-G-M] Chap. II.C, [CO] Chap. 3 and 7, or [C-E].

B. CURVATURES: The geometric point of view

As we shall see in a minute, there are several notions of curvature. These Riemannian invariants are very difficult to grasp and we will meet them under various circumstances. We first give definitions of a geometric flavor. (See [B-G-M] Chap. II.D and E, [CO] Chap. 4 and 8 or [C-E], [B-C]).

14. Sectional curvature. Let $x$ be a point in $(M,g)$ and let $P$ be a (2-dimensional) plane in $T_xM$. We call $C_P(r)$ the image, under the exponential map $\exp_x$, of the circle centered at $0$, with radius $r$, in $P$. This is a curve in $(M,g)$, whose length we call $\ell_p(r)$ (if $(M,g)$ is not complete the map $\exp_x$ might not be defined on the whole of $T_xM$, but it is always defined on a small ball centered at $0$ in $T_xM$).

It turns out that one can prove the following estimate

$$(15) \qquad \ell_p(r) = 2\pi r\left(1 - \frac{r^2}{6}\sigma(P) + 0(r^3)\right)$$

as $r$ goes to zero ([B-G-M] Chap. II.E.III).

The number $\sigma(P)$ which appears in (15) is called the sectional curvature of the 2-plane $P$ at $x$. This defines a function on the Grassmannian $G_{m,2}(T_xM)$ (the set of 2-planes in $T_xM$) and, when $x$ varies, a function on $G_{m,2}(M)$ the Grassmannian bundle over $M$ (see [NN]). In dimension 2, this is only a function on $(M,g)$. When $(M,g)$ is a surface in $\mathbb{R}^3$, with induced metric, the sectional curvature coincides with the Gaussian curvature of the surface (product of theprincipal curvatures), see [HF]. When $\dim M$ is bigger than 2, this is a much more complicated object.

16. Comments. The fact that $\ell_p(r) \sim 2\pi r$ as $r$ goes to zero, means that a Riemannian manifold looks like Euclidean space in the

small. The fact that there is no second order term in (15) comes from the fact that in a "good" coordinate system centered at $p$ in $M\{x_1,\ldots,x_n\}$, (namely the one given by $\exp_p$) one has $g_{ij}(0) = \delta_{ij}$ and $\frac{\partial g_{ij}}{\partial x_k}(0) = 0$.

Local calculations show that the curvature involves second order derivatives of the metric.

## 17. Examples.

(a) As is easily seen (see Fig. 1), in the case of the canonical 2-sphere (or more generally n-sphere) we have

$$\ell_p(r) = 2\pi \sin r = 2\pi r\left(1 - \frac{r^2}{6} + O(r^3)\right)$$

which shows that $\sigma \equiv 1$ on $(S^n, \mathrm{can})$;

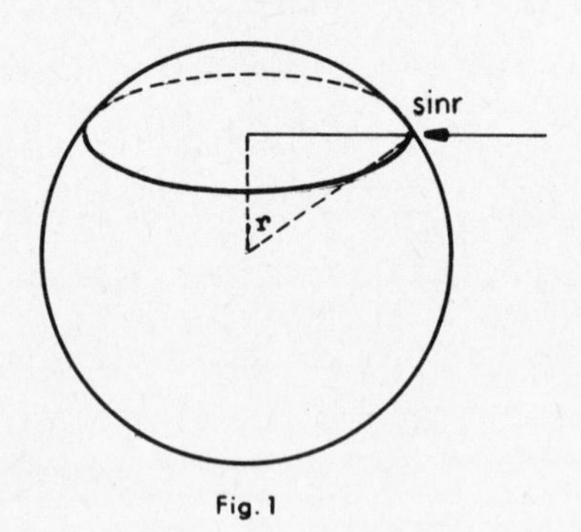


Fig. 1

(b) It is even easier to see that $\sigma \equiv 0$ on $(\mathbb{R}^n, \mathrm{can}), (T^n_G, \mathrm{can}/G)$;

(c) Exercise. Consider the Riemannian manifold $(H^n, \mathrm{can})$ given in Example 3(a). Show that the geodesics issued from $0$ are the rays issued from $0$ (Hint: use the differential equation and the fact that the image of a geodesic by an isometry again is a geodesic). Compute the length of the curve $t \to (tr, 0)$ where $r < 2$. Show that for any 2-plane $P$ in $T_oH^n$, $\ell_p(r) = 2\pi \,\mathrm{sh} r$. Conclude that $\sigma(P) = -1$.

In fact one can show that given any two points $x, y$ in $(H^n, \mathrm{can})$ there exists an isometry $f$ of $(H^n, \mathrm{can})$ such that

$f(x) = y$. It follows that $\sigma \equiv -1$ for $(H^n, can)$.

18. Remarks.

By adjusting the definitions of $(S^n, can)$ and $(H^n, can)$ (scaling) one can easily construct the simply connected Riemannian manifolds $(\$^n_k, can)$ whose sectional curvature is constant and equal to $k$ (any real number). They are called space forms. If $k > 0$ $(\$^n_k, can)$ is homothetic to $(S^n, can)$; if $k = 0$ $(\$^n_o, can)$ is just $(\mathbb{R}^n, can)$; if $k < 0$ $(\$^n_k, can)$ is homothetic to $(H^n, can)$ ([CO] Chap. 8).

The sectional curvature is a very strong invariant. Any Riemannian manifold $(M,g)$ whose sectional curvature is constant equal to $k$, is locally isometric to $(\$^n_k, can)$: given any point $x$ in $(M,g)$, there exists a neighborhood of $x$ which is isometric to a neighborhood of a point in $(\$^n_k, can)$. This local property is in fact global when $M$ is simply-connected: a simply-connected complete Riemannian manifold $(M,g)$ with constant sectional curvature equal to $k$ is isometric to $(\$^n_k, can)$ (These results are known as E. Cartan's theorems, [B-G-M] Chap. II.E.III).

19. An Example.

We denote by $\mathbb{C}P^n$ the complex projective space (complex lines in $\mathbb{C}^{n+1}$) i.e. $\mathbb{C}^{n+1}\setminus\{0\}/\mathbb{C}^*$.

We can identify $\mathbb{C}^{n+1}$ and $\mathbb{R}^{2n+2}$, so that the unit sphere $S^{2n+1}$ is the set $\{(z_o, z_1, \ldots, z_n) \in \mathbb{C}^{n+1} \mid |z_o|^2 + \ldots + |z_n|^2 = 1\}$. The circle $S^1$ acts on $S^{2n+1}$ by $e^{it}.(z_o, \ldots, z_n) = (z_o e^{it}, \ldots, z_n e^{it})$. It is easy to see that $\mathbb{C}P^n$ is diffeomorphic to $S^{2n+1}/S^1$. Here we view $\mathbb{C}P^n$ as a real manifold. Let $p: S^{2n+1} \to \mathbb{C}P^n$ the projection map. Given $x$ in $S^{2n+1}$, we let $H_x$ denote the space $\mathbb{C}^{n+1}$-orthogonal to the complex line $\mathbb{C}x$ of $\mathbb{C}^{n+1}$ ($\dim_{\mathbb{C}} H_x = n$). The tangent map to $p$ defines an isomorphism $p^T$ from $H_x$ onto $T_{p(x)}\mathbb{C}P^n$. We define a metric $g$ on $\mathbb{C}P^n$ as follows. Given any vector $X$ in $H_x$ we let $g(p^T(X),p^T(X)) = |X|^2$ (norm in $\mathbb{C}^{n+1}$).

This metric $g$ turns the map $p$ into a Riemannian submersion $p: (S^{2n+1},can) \to (\mathbb{C}P^n,g)$ with fibers $S^1$. This means that $p$ is a (differentiable) submersion and that the tangent map $T_x p$ is an isometry from the orthogonal to the tangent space to the fiber at $x$ onto the tangent space to $\mathbb{C}P^n$ at $p(x)$. The geodesics are easily seen to be the images under $p$ of the great circles of $S^{2n+1}$ which are orthogonal to the fibers of $p$. One can show that the sectional curvature of a 2-plane $P$ in $T\mathbb{C}P^n$ lies between 1 and 4 ([BS] Chap. 3).

20. The sectional curvature measures how geodesics diverge from one another (this should be at least intuitive from the very definition of sectional curvature: see nº (15)).

Let us make this statement more precise. Let us consider a point $x$ in the manifold $(M,g)$, and two geodesics $c_1(t)$, $c_2(t)$ such that

$$(21)\quad \begin{cases} \text{(i)} & c_1(0) = c_2(0) = x; \\ \text{(ii)} & c_1'(0) \text{ and } c_2'(0) \text{ form an angle } A; \\ \text{(iii)} & c_1(t),\ c_2(t) \text{ are parametrized by arc-length.} \end{cases}$$

22. Since geodesics are locally length minimizing, it follows that for $t$ small enough, the Riemannian distance $d(x,c_i(t))$ is just equal to $t$. Let us consider the geodesic triangle $\{x,c_1(t_1),c_2(t_2)\}$ whose sides are the minimizing geodesics between the vertices ($t_1,t_2$ are assumed to be small). Let us call $T(t_1,t_2)$ the length of the side from $c_1(t_1)$ to $c_2(t_2)$.

Let us now call $\{\bar{x},\bar{c}_1(t_1),\bar{c}_2(t_2)\}$ a geodesic triangle on $(\$^n_k,\mathrm{can})$ where $\bar{x}$ is in $(\$^n_k,\mathrm{can})$ and where $\bar{c}_i(t)$ are geodesics satisfying assumptions analogous to (21). Let us call $\bar{T}_k(t_1,t_2)$ the length of the side from $\bar{c}_1(t_1)$ to $\bar{c}_2(t_2)$ (again $t_1,t_2$ are assumed to be small enough). See Fig. 2.

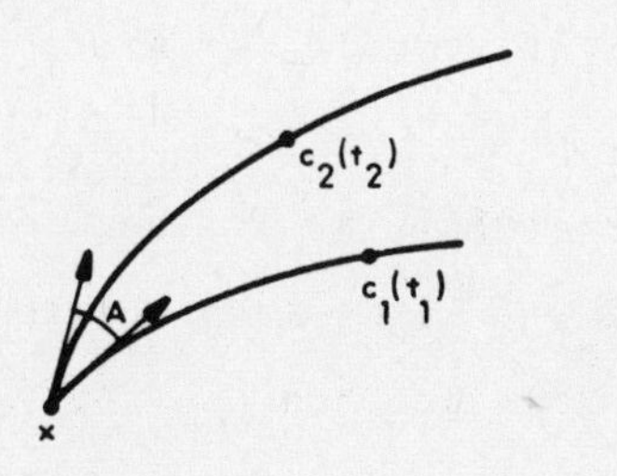


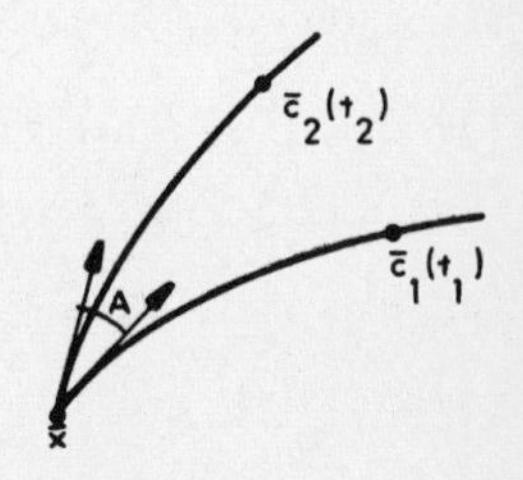


Fig. 2

23. <u>Definition</u>

We say that the sectional curvature $\sigma = \sigma(M,g;.)$ of $(M,g)$ is <u>bigger than</u> $k$ if for any 2-plane $P$ in $TM$ (i.e. any point $x$ in $M$, any 2-plane $P_x$ in $T_xM$), $\sigma(P) > k$ (one defines $\geq$, $<$, $\leq$ similarly).

We then have the following <u>comparison theorem</u>

24. <u>Theorem</u> (i) (Rauch)

<u>Let</u> $(M,g)$ <u>be a Riemannian manifold whose sectional curvature satisfies</u> $\sigma \geq k$ (<u>resp</u>. $\sigma \leq k$) <u>then, with the notations of</u> nº 22, <u>we have</u>

$$T(t_1,t_2) \leq \bar{T}_k(t_1,t_2) \quad (\text{resp. } T(t_1,t_2) \geq \bar{T}_k(t_1,t_2)),$$

*for all geodesic triangles constructed an in* n° 22, *for* $t_1, t_2$ *small enough*.

(ii) (Rauch-Alexandrov-Toponogov)

*If* $\sigma \leq k$ *then* $T(t_1,t_2) \geq \bar{T}_k(t_1,t_2)$ *holds for all geodesic triangles constructed as in* n° 22 (*whatever the size of* $t_1, t_2$).

This theorem is difficult; see [C-E] Chap. 2 for more precise statements and proofs, see also [SI] for a survey.

25. *Ricci curvature*. Given a point $x$ in the Riemannian manifold $(M,g)$, it is easy to show that $\exp_x$ is a local diffeomorphism from a neighborhood of $0$ in $T_xM$ onto a neighborhood of $x$ in $M$. The pulled-back measure $\exp_x^*(v_g)$ has a density with respect to the Lebesgue measure in $T_xM$. Using polar coordinates in $T_xM$, $(t,u) \in \in \mathbb{R}_+^\bullet \times S^{n-1}$, we can write $\exp_x^*(v_g) = \theta_x(t,u)dt\, du$, at least for $t$ small enough. More precisely, let $\phi_x : \mathbb{R}_+^\bullet \times S^{n-1} \to M$ be defined by $\phi_x(t,u) = \exp_x(tu)$, where $S^{n-1}$ is the unit sphere in $(T_xM,g_x)$. Then $\phi_x^* v_g = \theta_x(t,u)dt\, du$. The following expansion holds

$$\theta_x(t,u) = t^{n-1}\{1 - \frac{t^2}{6} r_x(u,u) + 0(t^3)\} \quad (n=\dim M), \tag{26}$$

as $t$ goes to zero ([B-G-M] Chap. II.E.III).

Here $r_x(u,u)$ is a quadratic form in $u$, whose associated symmetric bilinear form is called the *Ricci curvature* at $x$ (we shall often forget the index $x$ in $\theta_x, r_x$). The function $\theta_x(t,u)$ is the density of the Riemannian measure viewed through the map $\phi_x$. See also [BS] Chap. 6 p. 154.

27. *Comments*

The fact that $\theta_x(t,u) \sim t^{n-1}$ as $t$ goes to zero means that the Riemannian measure is asymptotically Euclidean. The Ricci curvature measures how the Riemannian measure differs from the Euclidean Lebesgue measure, at least infinitesimally. Let $u$ be a unit vector in $T_xM$ and let $\{u,e_2,\ldots,e_n\}$ be an orthonormal basis of $T_xM$. Let $P_i = [u,e_i]$ denote the 2-plane spanned by the vectors $u$ and $e_i$ in $T_xM$. We then have

$$r_x(u,u) = \sum_{i=2}^{n} \sigma(P_i). \tag{28}$$

It follows from Formula (28) that the Ricci curvature of $(S^n_k,\text{can})$ satisfies $r(u,u) = (n-1)k$, for any unit tangent vector $u$ or equivalently $r = (n-1)k$ can.

Formula (28) also shows that an assumption on the Ricci curvature is weaker than an assumption on the sectional curvature.

Let us mention the following important theorem which relates the Ricci curvature and the diameter of the manifold $(M,g)$.

29. Theorem (Myers)

*Let* $(M,g)$ *be a complete Riemannian manifold whose Ricci curvature satisfies* $r(u,u) \geq (n-1)k > 0$ *for any unit tangent vector* $u$.

*Then the diameter of* $(M,g)$ *satisfies*

$$\text{Diam}(M,g) \leq \pi/\sqrt{k},$$

*and hence* $(M,g)$ *is compact. Furthermore the fundamental group* $\pi_1(M)$ *is finite*. ([CO] Chap. 9, [C-E] Chap. 1).

The following *comparison theorem* will be of utmost importance in the sequel.

30. Theorem

*Let* $(M,g)$ *be a* n-*dimensional Riemannian manifold whose*

Ricci curvature $r$ satisfies $r(u,u) \geq (n-1)k$ for any unit tangent vector $u$, $k$ any real number. Let $B_k^n(t)$ be any geodesic ball with radius $t$ in $(\$_k^n, \mathrm{can})$ (all such balls are isometric). Then

(i) (Bishop) For any point $x$ in $M$ and $t$ in $\mathbb{R}_+$, $\mathrm{Vol}(B(x,t)) \leq \mathrm{Vol}(B_k^n(t))$ ($B(x,t)$ is the geodesic ball with radius $t$ and center $x$ in $(M,g)$);

(ii) (Gromov) The function

$$t \to \mathrm{Vol}(B(x,t))/\mathrm{Vol}(B_k^n(t))$$

is non-increasing. ([B-C] or [GV1]).

31. Caution. Reverse inequalities when $r(u,u) \leq (n-1)k$ DO NOT HOLD (exept when $n = 2$ or $n = 3$).

32. Curvature versus scaling. Let $(M,g)$ be a Riemannian manifold and let $c: [0,1] \to M$ be a curve. If the length of $c$ in $(M,g)$ is equal to $L$, then the length of $c$ in $(M,ag)$ (where $a > 0$) is $\sqrt{a}L$. If we denote by $\sigma(M,g)$ the sectional curvature of the Riemannian manifold $(M,g)$, it follows from (15) that $\sigma(M,ag) = a^{-1}\sigma(M,g)$ $(a > 0)$; the sectional curvature is Riemannian invariant of weight $-1$ (see nº 6). The products $\mathrm{Diam}(M,g)^2\sigma(M,g)$, $\mathrm{Vol}(M,g)^{2/n}\sigma(M,g)$, are therefore Riemannian invariants of weight $0$.

Let $r(M,g) = r$ (resp. $r_{min}(M,g)$) denote the Ricci curvature (resp. $\inf\{\frac{r(u,u)}{g(u,u)};\ u \in TM,\ u \neq 0\}$) of the Riemannian manifold $(M,g)$.

33. Exercise. Show that $r(M,g)$ is a Riemannian invariant of weight $0$, and that $r_{min}(M,g)$ is a Riemannian invariant of weight $-1$ (Hint: use formula (28) and the fact that $r(M,g)$ is a bilinear form on $TM$).

## C. THE COVARIANT DERIVATIVE

34. Given a manifold $M$ and a function $f: M \to \mathbb{R}$, one can define the differential of $f$, $df$ as follows. Let $p$ be a point in $M$ and $U$ a tangent vector in $T_pM$. Let $\{x_1,\dots,x_n\}$ be a local coordinate system centered at $p$. The function $f$ can then be viewed as a function $f(x_1,\dots,x_n)$. We let

$$df_p(U) = \sum_{i=1}^{n} \frac{\partial f}{\partial x_i}(0)\, u_i,$$

if $U = \sum_{i=1}^{n} u_i \frac{\partial}{\partial x_i}$. It is easy to prove that $df$ is invariantly defined on $TM$. A straightforward computation shows that $\left(\frac{\partial^2 f}{\partial x_i \partial x_j}(0)\right)$ $1 \le i,j \le n$, does not define an invariant object on $M$, unless $df_p = 0$.

One of the main features of Riemannian geometry is that to a Riemannian metric $g$ on a manifold $M$, is naturally attached an intrinsic notion of derivation.

Let $\mathfrak{A}(M)$ denote the vector space of $C^\infty$ vector-fields on $M$.

35. Theorem and Definition.

*Let $(M,g)$ be a Riemannian manifold. There is a unique map*

$$D: \mathfrak{A}(M)\times\mathfrak{A}(M) \to \mathfrak{A}(M)$$

$$(X,Y) \to D_X Y,$$

*with the following properties (for any $X,Y,Z$ in $\mathfrak{A}(M)$ and $f$ in $C^\infty(M)$)*

(i) $X.g(Y,Z) = g(D_X Y,Z) + g(Y,D_X Z)$;

(ii) $D_X Y - D_Y X = [X,Y]$;

(iii) $D$ is $\mathbb{R}$-bilinear;

(iv) $D_{(fX)}Y = f\, D_X Y$;

(v) $D_X(fY) = (X.f)Y + f\, D_X Y$.

<u>This map</u> $D$ <u>is called the</u> Levi-Civita <u>connection of the Riemannian manifold</u> $(M,g)$. ([B-G-M], II.B. I. 4, p. 24).

36. <u>Exercise</u>. Using Property (iv) above, show that $(D_X Y)(x)$ depends only on the value $X_x$ of the vector-field $X$ at $x$.

<u>Caution</u>: the same property does not hold for $Y$, see nº 39(3).

37. <u>Metrics and Connections on tensor products</u>. Let $\{e_i\}_1^n$ be an orthonormal basis of $T_x M$. Let $\{e_i^*\}_1^n$ be the dual basis in $T_x^* M$, the dual space of $T_x M$. We extend $g_x$ to a scalar product $g_x^*$ on $T_x^* M$ such that the basis $\{e_i^*\}_1^n$ be orthonormal.

<u>Exercise</u>. Show that the matrix $(g_{ij}^*)$ of $g^*$ in a local coordinate system $\{x_1,\ldots,x_n\}$ is $(g_{ij})^{-1}$, the inverse of the matrix $(g_{ij})$ of the metric $g$.

We extend $g_x$ to a scalar product on $\otimes^p T_x M \otimes^q T_x^* M$ by taking an orthonormal basis $\{e_i\}_1^n$ of $T_x M$ and by requiring that the natural basis of $\otimes^p T_x M \otimes^q T_x^* M$ deduced from $\{e_i\}_1^n$ be orthonormal.

We can also extend the Levi-Civita connection on tensors. For this purpose we require that Leibnitz rule be true, e.g. if $U, V, X$ are sections of $TM$ and if $W$ is a section of $T^* M$, we let (using the same symbol $D$ for the extension of the connection)

$$(38) \qquad \begin{cases} \text{(i)} & D_X(U \otimes V) = (D_X U) \otimes V + U \otimes (D_X V), \\ \\ \text{(ii)} & X.(W(U)) = (D_X W)(U) + W(D_X U). \end{cases}$$

We also define $D_X f$, for $f$ in $C^\infty(M)$ as $X.f$. This extension of the Levi-Civita connection satisfies properties similar to those of Theorem 35.

39. Exercises.

The "musical" isomorphisms $TM \overset{\flat}{\underset{\#}{\rightleftarrows}} T^*M$ are defined as follows: for $u$ in $TM$ and $f$ in $T^*M$, we let

$$g(u,f^\#) = f(u), \text{ and}$$
$$u^\flat = g(u,.).$$

(1) Show that for all $X,Y$ in $\mathfrak{A}(M)$,

$$D_X(Y^\flat) = (D_X Y)^\flat.$$

(2) Show that for all $X$ in $\mathfrak{A}(M)$,

$$D_X g = 0 \quad (\text{view } g \text{ as a section of } \otimes^2 T^*M).$$

(3) Let $p$ be a point in $(M,g)$ and let $(U,F)$ be a chart centered at $p$, i.e. $U$ is an open set in $M$, containing $p$ and $F\colon U \to F(U) \subset \mathbb{R}^n$ is a diffeomorphism, with $F(p) = 0$. Let $\{x_1,\ldots,x_n\}$ be local coordinates on $U$. Let $X,Y$ be vector-fields on $M$, whose expressions in $(U,F)$ are $\sum_{i=1}^n X_i(x_1,\ldots,x_n)\frac{\partial}{\partial x_i}$, $\sum_{i=1}^n Y_i(x_1,\ldots,x_n)\frac{\partial}{\partial x_i}$. Show that there exist $C^\infty$ functions on $U$, $\Gamma^i_{jk}$, such that $D_X Y$ is represented by the vector-field

$$\sum_{i=1}^n \{\sum_{j=1}^n X_j(x_1,\ldots,x_n)[\frac{\partial Y_i}{\partial x_j}(x_1,\ldots,x_n) + \sum_{k=1}^n \Gamma^i_{jk}(x_1,\ldots,x_n)Y_k(x_1,\ldots,x_n)]\}\frac{\partial}{\partial x_i}.$$

Compare with Exercise 36. The coefficients $\Gamma^i_{jk}$ are called the Christoffel symbols of the metric $g$.

(4) Let $p$ in $M$, and $u$ in $T_pM$. Let $Y_1, Y_2$ be vector-fields on $M$ in a neighborhood of $p$. Let $c: ]-a,a[ \to M$ be a $C^\infty$-curve such that $c(0) = p$ and $c'(0) = u$. Assume that for all $t$ in $]-a,a[$, $Y_1(c(t)) = Y_2(c(t))$. Show that $D_uY_1 = D_uY_2$.

(5) Find the Levi-Civita connection on $(\mathbb{R}^n, \text{can})$ (Hint: compare it with the usual derivation of a vector-field).

(6) Let $D$ be the Levi-Civita connection on $(S^n, \text{can})$, and let $\tilde{D}$ be the Levi-Civita connection on $(\mathbb{R}^{n+1}, \text{can})$. Let $X,Y$ be vector fields on $S^n$. Show that they can be extended to local vector fields on $\mathbb{R}^{n+1}$, $\tilde{X}, \tilde{Y}$, and that for any $x$ in $S^n$, $(D_XY)(x)$ is the orthogonal projection of $(\tilde{D}_{\tilde{X}}\tilde{Y})(x)$ on $T_xS^n$ (Hint: use Theorem 35 and Exercise 39(4)).

40. Let $c: ]-a,a[ \to M$ be a $C^\infty$-curve in $M$. A <u>vector-field along</u> $c$ is a $C^\infty$ map $X: ]-a,a[ \to TM$ such that $X(t)$ is in $T_{c(t)}M$ for all $t$ in $]-a,a[$. It follows from Exercise 39(4) that one can define $D_{c'(t)}X$. We say that <u>the vector-field</u> $X$ <u>is parallel along the curve</u> $c$ if $D_{c'(t)}X = 0$ for all $t$.

In a local coordinate system $\{x_1,\ldots,x_n\}$ near the point $c(0)$, we can write $c(t)$ as the curve $(A_1(t),\ldots,A_n(t))$ in $\mathbb{R}^n$, and $c'(t)$ as the vector $(A_1'(t),\ldots,A_n'(t))$ in $\mathbb{R}^n$. Let $\{X_i(t)\}_1^n$ be the coordinates of $X(t)$ in this coordinate system. According to Exercises 39(3) and (4) the condition $D_{c'(t)}X = 0$ can be written as

$$\sum_{j=1}^n A_j'(t)\left[\frac{\partial \tilde{X}_i}{\partial x_j}(A_1(t),\ldots,A_n(t)) + \sum_{k=1}^n \Gamma_{jk}^i\,\tilde{X}_k(A_1(t),\ldots,A_n(t))\right] = 0$$

or

$$\frac{d}{dt}X_i(t) + \sum_{j,k=1}^n A_j'(t)\Gamma_{jk}^i\,X_k(t) = 0, \qquad 1 \le i \le n \tag{41}$$

(where $\tilde{X}$ is an extension of $X$ in a neighborhood of $c(0)$ and $\tilde{X}_k(A_1(t),\ldots,A_n(t)) \equiv X_k(t)$).

Equation (41) is a system of ordinary differential equations on $]-a,a[$. Given a vector $u$ in $T_{c(0)}M$, one can therefore find a vector field $X$ along $c$, such that $D_{c'(t)}X = 0$. A <u>geodesic</u> is a curve $c(t)$, whose tangent vector $c'(t)$ is parallel along $c(t)$, $D_{c'(t)}c'(t) = 0$.

42. <u>Remark</u>. The map $u \to X(t)$ where $X$ is the parallel vector-field along $c$, such that $X(0) = u$, is called the <u>parallel translation</u> along $c$. This very important notion can be generalized to tensors on $M$.

<u>Exercise</u>. Let $c(t)$ be a curve in $(\mathbb{R}^n, \mathrm{can})$ and let $u$ be a vector at $c(0)$. Find the parallel vector-field $X(t)$ along $c$, such that $X(0) = u$.

For § C see [B-G-M] Chap. II. B, [CO] Chap. 2, or [C-E].

## D. <u>CURVATURES</u>: <u>The analytic point of view</u>.

43. The <u>curvature tensor</u> $R$ of the Riemannian manifold $(M,g)$ is defined as follows. Given $X,Y$ in $\mathfrak{X}(M)$ one defines the map $R(X,Y)$ from $\mathfrak{X}(M)$ to $\mathfrak{X}(M)$ by $R(X,Y) = [D_X, D_Y] - D_{[X,Y]}$, i.e. for any $U$ in $\mathfrak{X}(M)$,

$$R(X,Y)U = D_X(D_Y U) - D_Y(D_X U) - D_{[X,Y]}U.$$

<u>Caution</u>. The sign of $R$ may differ from one book to another.

44. <u>Properties</u>.

(i) <u>For any</u> $X,Y$ <u>in</u> $\mathfrak{X}(M)$, $R(X,Y) + R(Y,X) = 0$;

(ii) <u>The map from</u> $\mathfrak{X}(M)^3$ <u>to</u> $\mathfrak{X}(M)$, <u>which associates</u> $R(X,Y)U$ <u>to</u> $(X,Y,U)$ <u>is in fact a map from</u> $(T_xM)^3$ <u>to</u> $T_xM$ <u>i.e.</u>

$(R(X,Y)U(x)$ <u>depends only on the values of the vector fields</u> $X_x, Y_x, U_x$ <u>at</u> $x$ (we say that it is a <u>tensor</u>).

Proof. Use Theorem 35 to show that for any $f,g,h$, in $C^\infty(M)$ and any $X,Y,U$ in $\mathfrak{U}(M)$, $R(fX,gY)(hU) = fgh\,R(X,Y)U$ (i.e. that $R$ is $C^\infty(M)$ -3-linear) ■

45. <u>Definitions</u>. (Caution with sign conventions)

Let us define the tensor $R(X,Y;U,V)$ by

$$R(X,Y;U,V) = g(R(X,Y)V,U). \quad \text{Then}$$

(i) If $P$ is a 2-plane spanned by $\{X,Y\}$, we define the <u>sectional curvature</u> $\sigma$ of $(M,g)$ on the 2-plane $P$ by $\sigma(P) = R(X,Y;X,Y)/g(X\wedge Y, X\wedge Y)$, i.e. $\sigma(P) = R(e,f;e,f)$ if $\{e,f\}$ is an orthonormal basis of the 2-plane $P$;

(ii) The <u>Ricci curvature</u> of $(M,g)$ is defined by

$$r(X,X) = \sum_{i=1}^{n} R(X,e_i;X,e_i),$$

for any vector $X$ in $T_xM$, where $\{e_i\}_{1,\ldots,n}$ is any orthonormal basis in $T_xM$;

(iii) The <u>scalar curvature</u> $u$ of the Riemannian manifold $(M,g)$ at $x$ is defined by

$$u(x) = \sum_{i,j=1}^{n} R(e_i,e_j;e_i,e_j) = \sum_{j=1}^{n} r(e_j,e_j),$$

where $\{e_i\}$ is any orthonormal basis of $T_xM$. The scalar curvature is a function on $(M,g)$.

46. <u>Claim</u>. These definitions coincide with those given in nº 14 and 25.

47. <u>Notations</u>.

In order to make things explicit without statements, we will often use the following obvious notations Sect, Sect(M,g); Ricci, Ricci(M,g); Scal, Scal(M,g). (see [B-G-M] Chap. II, [CO] Chap. 4).

48. Let $f$ be a $C^\infty$ function defined on the Riemannian manifold $(M,g)$. Let $X,Y$ be two vector-fields on $M$. We denote by $Ddf(X;Y)$ the one-form $D_X(df)$ evaluated on the vector-field $Y$. The $\mathbb{R}$-bilinear map $Ddf$ is called the Hessian of $f$ and denoted by Hess $f$ (with respect to the Riemannian metric $g$). According to nº (38) we have.

(49) $$\text{Hess } f(X,Y) = X.(df(Y)) - df(D_X Y).$$

50. Proposition.

The Hessian of a $C^\infty$ function $f$, Hess $f$, is a symmetric two tensor i.e.

(i) $(\text{Hess } f(X,Y))(x)$ depends only on $X_x$ and $Y_x$;

(ii) $\text{Hess } f(X,Y) = \text{Hess } f(Y,X)$.

Proof. Use the fact that being a tensor is equivalent to $C^\infty(M)$-linearity and Theorem 35. ■

This proposition answers the question which was raised in nº 34, and generalizes for 2nd order derivatives the well-known Schwarz theorem on functions of several variables. An important fact in Riemannian geometry is that Schwarz theorem no longer holds for higher order derivatives.

Let $f$ be a $C^\infty$ function on $M$ and let $X,Y,Z$ be three vector fields on $M$. The following lemma holds.

51. <u>Lemma</u>.

$$(D_X(D_Y df))(Z) - (D_Y(D_X df))(Z) - (D_{[X,Y]}df)(Z) = -R(X,Y;df^{\#},Z).$$

<u>Proof</u>. Write (using nº (38) the first term in the left-hand side as

$$\begin{aligned}(D_X(D_Y df))(Z) &= X.((D_Y df)(Z)) - (D_Y df)(D_X Z)\\ &= X.[Y.(df(Z)) - df(D_Y Z)] - Y.(df(D_X Z)) + df(D_Y(D_X Z)),\end{aligned}$$

and a similar expression for the second term. This gives $([D_X,D_Y]df)(Z) = [X,Y].(df(Z)) - df([D_X,D_Y]Z)$. Then use the definition of the curvature tensor to conclude that

$$\begin{aligned}([D_X,D_Y]df)(Z) - (D_{[X,Y]}df)(Z) &= -df(R(X,Y)Z)\\ &= -R(X,Y;df^{\#},Z),\end{aligned}$$

using nº 45 and nº 39. ∎

52. Take an orthonormal basis $\{e_i\}_1^n$ at $x$ in $M$. Using nº 40, one can extend $\{e_i\}_1^n$ to a local orthonormal frame $\{X_i\}_1^n$ such that $X_i(x) = e_i$ and $(D_{X_i}X_j)(x) = 0$; from Theorem 35(ii), we deduce that $[X_i,X_j](x) = 0$ and from Exercise 36, we conclude that $(D_{[X_i,X_j]}df)(x) = 0$. Finally, we deduce from Lemma 51 that

$$([D_{X_i},D_{X_j}]df)(X_k) = D^3f(X_i;X_j;X_k)-D^3f(X_j;X_i;X_k) = -R(X_i,X_j;df^{\#},X_k)$$

(the second equality is a notation), which shows that Schwarz theorem does not hold for derivatives of order 3, unless $R=0$. We can view the <u>curvature</u> as <u>an obstruction to commuting derivatives</u>.

53. The vector-field $df^{\#}$, dual to the 1-form $df$, which appears in Lemma 51 is called the <u>gradient</u> of $f$; it depends on the Riemannian metric $g$ whereas $df$ does not.

Further references for Chapter II: Introduction to Riemannian geometry: [CL] Chap. III, [MR], [SI]; Riemannian geometry: [B-C], [B-G-M], [C-E], [CO], [KG], [SK].

# CHAPTER III

## THE LAPLACIAN AND RELATED TOPICS

ALL RIEMANNIAN MANIFOLDS ARE ASSUMED TO BE SMOOTH, CONNECTED AND COMPLETE

Unless otherwise stated, vector-fields, forms, functions... will also be assumed smooth.

This chapter is mainly devoted to the Laplace-Beltrami operator (or Laplacian) acting on $C^\infty$-functions on a Riemannian manifold $(M,g)$.

### A. STARRING: The Laplacian and the Rayleigh quotient.

1. Let $(M,g)$ be a Riemannian manifold, with Levi-Civita connection $D$. Given a smooth vector-field $X$ on $M$, one defines the divergence of $X$ with respect to the Riemannian metric $g$, as the function $\mathrm{Div}_g X$ (or simply $\mathrm{Div}\, X$) defined by

(2) $$(\mathrm{Div}_g X)(x) = \mathrm{Trace}\,\{u \to D_u X\}\,(x),$$

where the trace of the endomorphism $u \to D_u X$ is taken in $T_x M$. Given an orthonormal basis $\{e_1,\dots,e_n\}$ of $T_x M$ $(n = \dim M)$, one can also write

(3) $$(\mathrm{Div}_g X)(x) = \sum_{i=1}^{n} g(D_i X, e_i).$$

*Note*. When $\{e_1,\ldots,e_n\}$ is an orthonormal basis we use the *notation* $D_i$ instead of $D_{e_i}$.

4. The *Laplace-Beltrami operator* (or *Laplacian*) acting on $C^\infty$ functions is defined by the formula

$$\Delta^g f = -\mathrm{Div}_g(df^\#)$$

where $df^\#$ is the gradient of $f$ (see nº II.53). We shall also write $\Delta$ instead of $\Delta^g$ (*Notice our sign convention*: compare with III.11(c)).

The following proposition gives useful formulas for the Laplacian.

5. *Proposition*. *Let* $(M,g)$ *be an* n-*dimensional Riemannian manifold*, $D$ *its Levi-Civita connection and* $\Delta$ *its Laplacian. Let* $\{e_1,\ldots,e_n\}$ *be a local orthonormal frame near the point* $p$ *in* $M$. *Let* $f$ *be a* $C^\infty$ *function on* $M$. *The following formulas hold*

(i) $\Delta f(p) = -\mathrm{Trace\ Hess}\ f(p)$,

*the trace of the bilinear form* Hess $f(p)$ *in* $T_pM$ *or equivalently*

$$\Delta f(p) = -\sum_{i=1}^{n} Ddf(e_i(p);e_i(p))$$

(*see* nº II.48);

(ii) $\Delta f(p) = -\sum_{i=1}^{n} \{e_i.(e_i.f) - (D_ie_i).f\}(p)$;

(iii) *Let* $\{x_1,\ldots,x_n\}$ *be a local coordinate system centered at* $p$. *Let* $g_{ij}(x) = g(\frac{\partial}{\partial x_i}, \frac{\partial}{\partial x_j})$ *and* $v = \mathrm{Det}(g_{ij})^{1/2}$. *We denote by* $(g^{ij})$ *the inverse matrix* $(g_{ij})^{-1}$. *The local expression of the Laplacian is*

$$(\Delta f)(x_1,\dots,x_n) = -[v^{-1} \sum_{i,j=1}^{n} \frac{\partial}{\partial x_i}(g^{ij} v \frac{\partial f}{\partial x_j})](x_1,\dots,x_n);$$

(iv) *Let* $\{c_1(t),\dots,c_n(t)\}$ *denote geodesics such that* $c_i(0) = p$ *and* $c_i'(0) = e_i(p)$.

*The function* $\Delta f$ *can be calculated at* $p$ *by the following formula*

$$(\Delta f)(p) = - \sum_{i=1}^{n} \left.\frac{d^2}{dt^2}\right|_{t=0} (f \circ c_i)(t);$$

(v) *Let* $p$ *be a fixed point in* $M$ *and let* $h$ *be a* $C^\infty$ *function on* $\mathbb{R}_+^\bullet$. *For* $r = d(p,x)$ *small enough we assume that the function* $x \to h(d(p,x)) = f(x)$ *is* $C^\infty$ (*for* $x \neq p$, *near* $p$). *We can write* $x = \exp_p(ru)$ (*for* $r$ *in* $\mathbb{R}_+^\bullet$ *and* $u$ *in the unit sphere* $S^{n-1}$ *of* $T_pM$) *and* $\exp_p^*(v_g) = \theta(r,u)drdu$ (*see* II.25).

*The following formula holds (near* $p$)

$$(\Delta f)(x) = -h''(d(p,x)) - \frac{\theta'(r,u)}{\theta(r,u)} h'(d(p,x)),$$

*where* $h'$ *and* $h''$ *are derivatives of* $h$ *and* $\theta'(r,u) = \frac{\partial\theta}{\partial r}(r,u)$.

*Proof*. The assertions (i) and (ii) follow from the definition of $\Delta$ and from the definition of Hess $f$, see Exercise III. 39(1) and nº II. 48.

(iii) Let $V = vdx_1\dots dx_n$ be the local volume form which represents the measure $v_g$ in the local coordinate system. A classical result ([B-G-M] Chap. II.G) states that given a vector field $X$, the Lie derivative $\mathcal{L}_X V$ is just $\mathrm{Div}_g(X)V$ or equivalently, since $V$ is an $n$-form, $d(i_X V) = \mathrm{Div}_g(X)V$. Writing $X$ as $\sum_{i=1}^{n} X_i \frac{\partial}{\partial x_i}$ in the local coordinates, one can deduce that the local expression for $\mathrm{Div}_g X$ is

(6) $\qquad \mathrm{Div}_g X = v^{-1} \sum_{i=1}^{n} \frac{\partial}{\partial x_i}(vX_i).$

The local expression for $df$ is $df = \sum_{i=1}^{n} \frac{\partial f}{\partial x_i} dx_i$, so that using the duality between $TM$ and $T^*M$ induced by the metric $g$, we have

$$df^{\#} = \sum_{i,j=1}^{n} (g^{ij} \frac{\partial f}{\partial x_j}) \frac{\partial}{\partial x_i}.$$

The assertion (iii) follows from these computations.

(iv) The local frame $\{e_1,\ldots,e_n\}$ can be obtained from $\{e_1(p),\ldots,e_n(p)\}$ by parallel translation along the geodesics issued from $p$. In particular (see nº II.41) we can choose $e_i$ such that $e_i(c_i(t)) = c_i'(t)$. It follows from this choice of $\{e_i\}$ that

(i) $\quad (e_i.(e_i.f))(p) = \frac{d^2}{dt^2} f\circ c_i(t)\Big|_{t=0}$, and

(ii) $\quad (D_{e_i} e_i)(p) = 0$. It suffices to apply Assertion (ii) and the definition of $Ddf$ (Notice that the final result is independent of the choice of the local orthonormal frame $\{e_i\}$).

(v) See [B-G-M] Chap. II.G or [CL]. ∎

7. <u>Comments</u>. The definition of $\Delta^g$ given in nº 4 shows that $\Delta^g$ is invariantly defined on $(M,g)$ and that $\Delta^g$ is a Riemannian invariant. Proposition 5(iii) shows that $\Delta$ is a 2nd order linear differential operator whose leading terms are $-\sum_{i,j=1}^{n} g^{ij} \frac{\partial^2}{\partial x_i \partial x_j}$. The function $\xi \to g^*(\xi,\xi)$, whose expression in local coordinates is $\sum_{i,j=1}^{n} g^{ij} \xi_i \xi_j$ is well defined on $T^*M$ and is called the <u>principal symbol</u> of the operator $\Delta$. For $x$ in $M$

the principal symbol maps $T_x^*M$ into $\mathbb{R}_+$ by $\xi \to g_x^*(\xi,\xi)$. It follows that $\Delta$ is <u>elliptic</u> (see [G-T] and [NN] Chap. 3); this fact will be very important in the sequel. Proposition 5(v) shows that the Laplacian $\Delta$ is strongly related to the Ricci curvature through $\theta(r,u)$ (see nº II.(26)). We shall use this property later on. Some of the formulas in Proposition 5 can be deduced from the fact that $\Delta$ is the Friedrichs extension of the quadratic form $u \to \int |\nabla u|^2$ ([R-S] Vol. II).

8. Let $(M,g)$ be a <u>Riemannian manifold with boundary</u>. The boundary $\partial M$ of $M$ is a Riemannian manifold with the induced metric $g|\partial M$. We use the following notations

a) $\overset{\circ}{M} = M \backslash \partial M$ (the interior of $M$);

b) $a_g$ the Riemannian measure on $(\partial M, g|\partial M)$;

c) $\nu$ the unit normal vector-field on $\partial M$, pointing inward.

FROM NOW ON ALL RIEMANNIAN MANIFOLDS WILL BE ASSUMED TO BE <u>COMPACT</u> unless otherwise stated.

The following theorems are standard (see [LG] p.204)

9. <u>Divergence Theorem</u>. <u>Let</u> $X$ <u>be a</u> $C^\infty$ <u>vector-field on</u> $M$. <u>Then</u>

$$\int_M (\mathrm{Div}_g X)(x)dv_g(x) = -\int_{\partial M} g(X,\nu)(x)da_g(x) = -\int_{\partial M} \langle X,\nu\rangle(x)da_g(x).$$

10. <u>Green's Theorem</u>. <u>Let</u> $f,h$ <u>be</u> $C^\infty$ <u>functions on</u> $M$. <u>Then</u>

(i) $$\int_M \{h(x)\Delta f(x) - g(\nabla h,\nabla f)(x)\}dv_g(x) = \int_{\partial M} h(x)(\nu.f)(x)da_g(x)$$

(ii) $$\int_M \{h(x)\Delta f(x) - f(x)\Delta h(x)\}dv_g(x) =$$

$$= \int_{\partial M} \{h(x)(\nu.f)(x) - f(x)(\nu.h)(x)\}da_g(x)$$

where $\nabla f = df^{\#}$ is the gradient of $f$ with respect to the Riemannian metric $g$ on $M$ (see nº II.53).

11. Remarks.

(a) In the sequel, we will simply write, e.g. for (ii)

$$\int_M (h\Delta f - f\Delta h)dv_g = \int_{\partial M} \{h(\nu.f)-f(\nu.h)\}da_g.$$

(b) Both theorems are true under more general assumptions: $M$ could be non-compact, provided that the integrations be in fact performed on compact sets (e.g. $f$, and $X$ with compact supports...); one can also weaken the regularity assumptions on $X$, $h$, $f$ (e.g. Theorem 9 works for $X$ a $C^1$-vector-field...), or on $\partial M$ ($\partial M$ might only be piece-wise smooth).

(c) In order to make things clear let us insist that our Laplacian is written $-f''$ on $\mathbb{R}$ and that our normal $\nu$ points inward.

Before we go any further with the study of the Laplacian, let us introduce some basic objects (compare with Chapter I nº43ff).

12. We denote by $L^2(M,v_g)$ or simply $L^2(M)$, the space of measurable functions $f$ on $M$ such that $\int_M |f(x)|^2 dv_g < +\infty$. This

space is a Hilbert space with inner product $(f|h)_o = \int_M fh dv_g$, and norm $\|f\|_o = (f|f)_o^{1/2}$ (we shall mainly deal with real-valued functions; when dealing with complex-valued functions we shall use $(f|h)_o = \int_M f\bar{h} dv_g$ as inner product).

We denote by $C_o^\infty(M)$ the set of $C^\infty$ functions on $M$, with compact support in $\overset{\circ}{M}$.

We define a norm on $C^\infty(M)$ by

$$\|f\|_1 = \{\int_M |f(x)|^2 dv_g(x) + \int_M |df|^2(x) dv_g(x)\}^{1/2}$$

where $|df|^2(x)$ is the square of the norm of the 1-form $df(x)$ in $T_x^*M$ i.e. $|df|^2(x) = g_x^*(df(x),df(x)) = g_x(\nabla f(x),\nabla f(x))$.

We shall now use the notation $\langle .|.\rangle_x$ or $\langle .|.\rangle$ for the natural scalar products on tensor products above the point $x$ in $M$ (see nº II.37) and $\langle\langle .|.\rangle\rangle$ for the integrated inner product. For example

$$\langle df|df\rangle_x = g_x^*(df(x),df(x)), \quad \text{and}$$

$$\langle\langle df|df\rangle\rangle = \int_M \langle df|df\rangle_x \, dv_g(x).$$

The norm $\|.\|_1$ is associated with the inner product

$$(f|g)_1 = (f|g)_o + \langle\langle df|dg\rangle\rangle$$

on $C^\infty(M)$.

13. Let us recall that $C^\infty(M)$ and $C_o^\infty(M)$ are dense in $L^2(M,v_g)$ for the norm $\|.\|_o$. The closure of $C^\infty(M)$ (resp. $C_o^\infty(M)$) in $L^2(M,v_g)$ for the norm $\|.\|_1$ will be denoted by $H^1(M,g)$ (resp. $H_o^1(M,g)$) or

simply $H^1(M)$ (resp. $H_o^1(M)$). The following inclusions are continuous (with the natural norms)

$$H_o^1(M) \subset H^1(M) \subset L^2(M).$$

These spaces are called Sobolev spaces (see [G-T]).

Let us point out that whereas $L^2(M,v_g)$ only depends on the measure $v_g$ ($M$ being a fixed manifold), $H^1(M,g)$ and $H_o^1(M,g)$ depend on the Riemannian metric itself since $\langle\langle df|dg\rangle\rangle$ does.[1]

The elements in $H^1(M)$ are $L^2$-functions on $M$, with first derivatives (in the sense of distributions) in $L^2(M)$ (see [G-T] p. 142). If $\partial M = \emptyset$ then $H_o^1(M) \equiv H^1(M)$.

The following theorem is standard ([G-T] p. 160 or [NN] Chap. 3).

14. Theorem. *Let $(M,g)$ be a smooth compact Riemannian manifold with boundary (possibly empty). The inclusion maps*

(i) $(H_o^1(M,g),\|.\|_1) \to (L^2(M,v_g),\|.\|_o)$, *and*

(ii) $(H^1(M,g),\|.\|_1) \to (L^2(M,v_g),\|.\|_o)$

*are compact (or completely continuous): the image in $L^2$ of a bounded set in $H^1$ or $H_o^1$ is relatively compact.*

15. Remark. The theorem remains true under weaker assumptions on the regularity of $\partial M$; however the second assertion might be false if $\partial M$ is too irregular.

---

(1) Here, we consider $H^1$ or $H_o^1$ equipped with the norm $\|.\|_1$, and not only the spaces of functions, which do not depend on $g$ because $M$ is compact.

## B. EIGENVALUE PROBLEMS ON RIEMANNIAN MANIFOLDS I.

In these notes we shall be interested in the following eigenvalue problems:

(16C) $(M,g)$ is a compact Riemannian manifold without boundary, $\Delta u = \lambda u$ (<u>closed eigenvalue problem</u>);

(16D) $(M,g)$ is a compact Riemannian manifold with boundary,

$$\begin{cases} \Delta u = \lambda u & \text{in } \overset{\circ}{M}, \\ u = 0 & \text{on } \partial M, \end{cases} \qquad \text{(\underline{Dirichlet eigenvalue problem})};$$

(16N) $(M,g)$ is a compact Riemannian manifold with boundary,

$$\begin{cases} \Delta u = \lambda u & \text{in } \overset{\circ}{M}, \\ \nu . u = 0 & \text{on } \partial M, \end{cases} \qquad \text{(\underline{Neumann eigenvalue problem})},$$

where $\nu$ is the unit normal vector-field on $\partial M$, pointing inward (see nº 8).

This means that given the compact Riemannian manifold $(M,g)$, we look for all numbers $\lambda$ for which there exists a nontrivial solution $u$ in $C^\infty(M)$ of the (boundary value) Problem (16C), (16D) or (16N).

Notice that if $u$ is a nontrivial solution of one of the Problems (16), then the corresponding number $\lambda$ must be a nonnegative real number (apply Green's Theorem 10(i) with $h = f = u$).

17. The numbers $\lambda$ for which (16*) has a nontrivial solution $u$ in $C^\infty(M)$ are called the <u>eigenvalues</u> of problem (*), * = C,D or N. The corresponding functions $u$, called the <u>eigenfunctions</u> of problem (*) associated with the eigenvalue $\lambda$, form a vectorspace whose dimension is called the <u>multiplicity</u> of the eigenvalue $\lambda$.

18. Theorem. Let $(M,g)$ be a compact Riemannian manifold and let $(*)$ be one of the eigenvalue problems (C),(D) or (N) of nº 16.

(i) The set of eigenvalues of problem $(*)$ consists of an infinite sequence $(0\leq)\ \bar{\lambda}_1 < \bar{\lambda}_2 < \bar{\lambda}_3 < \ldots \uparrow +\infty$;

(ii) Each eigenvalue $\bar{\lambda}_i$ has finite multiplicity and the eigenspaces corresponding to distinct eigenvalues are $L^2(M,v_g)$ - orthogonal;

(iii) The direct sum of the eigenspaces $E(\bar{\lambda}_i)$ $i = 1,2,\ldots,$ is dense in $L^2(M,v_g)$ for the $L^2$-norm topology and dense in $C^k(M)$ for the uniform $C^k$-topology, $k = 0,1,2,\ldots$ .

19. Notations. From now on, we will list the eigenvalues of Problem $(*)$ as $(0\leq)\ \lambda_1 \leq \lambda_2 \leq \lambda_3 \leq \ldots \uparrow +\infty$ with each eigenvalue repeated a number of times equal to its multiplicity. If necessary we will write

$$\lambda_i,\ \lambda_i(M,g,*),\ \lambda_i(M,g) \text{ or } \lambda_i(*)$$

to point out the dependence on the manifold $(M,g)$, the eigenvalue problem $*$ which is considered, or both the manifold and the eigenvalue problem.

To the sequence $\lambda_1 \leq \lambda_2 \leq \lambda_3 \leq \ldots$ formed by the eigenvalues of Problem $(*)$, one can associate an orthonormal family $\phi_1,\phi_2,\ldots$ of eigenfunctions such that $\phi_i$ satisfies the eigenvalue Problem $(16*)$ with $\lambda = \lambda_i$, $* = C,D,N$.

The third assertion in Theorem 18 shows that the sequence $\{\phi_i\}_{i=1}^{\infty}$ is an orthonormal basis of $L^2(M,v_g)$. For any $f$ in $L^2(M)$, one can write

$$f = \sum_{i=1}^{\infty} (f|\phi_i)_o \phi_i \quad \text{in } L^2\text{-sense, and} \quad \|f\|_o^2 = \sum_{i=1}^{\infty} (f|\phi_i)_o^2 .$$

Let us sketch two possible proofs for Theorem 18.

20. Let $D_*$, $* = C,D,N$, denote the following subspaces of $C^\infty(M)$, which are dense in $L^2(M)$

$$D_C = C^\infty(M)$$

$$D_D = \{f \text{ in } C^\infty(M) \mid f = 0 \text{ on } \partial M\}$$

$$D_N = \{f \text{ in } C^\infty(M) \mid \nu . f = 0 \text{ on } \partial M\}.$$

In order to study the eigenvalue Problem (16*), one is led to view the Laplacian $\Delta$ as an unbounded operator in $L^2(M)$, with domain $D_*$. It follows from Green's Theorem 10 that $\Delta$ is

<u>Symmetric</u> i.e. for any $f,h$ in $D_*$,

$$(f|\Delta h)_o = (\Delta f|h)_o;$$

<u>Positive</u> i.e. for any $f$ in $D_*$,

$$(\Delta f|f)_o = \langle\langle \nabla f|\nabla f\rangle\rangle \geq 0.$$

A classical theorem in spectral theory ([R-S] Chap. X or [TR1] Part 3 and [TR2] Section 3) states that $(D_*,\Delta)$ has a unique extension $(\mathcal{E}_*,\Delta_*)$ as an unbounded self-adjoint operator in $L^2(M,v_g)$. The vector-space $\mathcal{E}_*$ is contained in the subset of functions in $L^2(M)$, whose derivatives (in the sense of distributions) up to order 2 are in $L^2(M)$. The operator $\Delta_*$ is then $\Delta$ viewed on $\mathcal{E}_*$ as a differential operator acting on distributions. It must be pointed out that $(\mathcal{E}_D,\Delta_D)$ and $(\mathcal{E}_N,\Delta_N)$ are quite different operators: they contain both the Laplacian $\Delta$ as a differential operator acting on distributions and the boundary conditions (Dirichlet or Neumann). The positivy of $\Delta$ implies that $(\mathcal{E}_*,\Delta_*)$ is a positive self-adjoint operator which in turn implies that the <u>spectrum</u> of $(\mathcal{E}_*,\Delta_*)$ is contained in $\mathbb{R}_+$. The compactness and regularity assumptions on

M imply that the inclusion $\mathcal{E}_* \to L^2(M)$ is compact. It follows that for $\lambda \notin \mathbb{R}_+$, the resolvent $(\Delta_* - \lambda)^{-1}$ is a compact operator in $L^2(M)$. Theorem 18 follows from the classical results on the spectral theory of compact operators and from the fact that the Laplacian $\Delta$ is an elliptic differential operator (for more details see [TR1] and [TR2] or [AN], [R-S] Vol. II).

21. Instead of looking at the Laplacian $\Delta$, one can consider the Dirichlet integral or the Rayleigh (-Ritz) quotient: for u in $C^\infty(M)$ we define

$$E(u) = \int_M |du|^2 v_g \qquad \text{(Dirichlet or energy integral);}$$

$$R(u) = \int_M |du|^2 v_g \Big/ \int_M u^2 v_g \qquad \text{(Rayleigh quotient),}$$

where $\int_M u^2 v_g \neq 0$. For motivations see Chap. I, where we pointed out the two points of view: operator vs quadratic form.

Both $E(u)$ and $R(u)$ are defined on $H^1_*(M)$ (see nº 13).

In order to prove Theorem 18, one considers the extrema of $R(u)$ on $H^1_*(M)$ or equivalently on $C^\infty_*(M)$, where

$$H^1_C(M) = H^1(M), \quad C^\infty_C(M) = C^\infty(M);$$
$$H^1_D(M) = H^1_o(M), \quad C^\infty_D(M) = C^\infty_o(M);$$
$$H^1_N(M) = H^1(M), \quad C^\infty_N(M) = C^\infty(M);$$

denote the sets of admissible functions (see Chapter I, nº I.9-11) respectively for the Closed, Dirichlet or Neumann Eigenvalue problems.

Let $\mu^*_1 = \inf\{R(u) : u \in H^1_*(M), \int_M u^2 \neq 0\}$.

This infimum exists because $R(u)$ is non-negative for all $u$. Let $\{u_n\}_1^\infty$ be a sequence in $H_*^1(M)$, normalized by $\int_M u_n^2 = 1$, such that $\mu_1^* \le R(u_n) \le \mu_1^* + \frac{1}{n}$.

From the definition of $R(u)$ we deduce that, for all $n$, $\|u_n\|_1^2 \le \mu_1^* + 2$. The sequence $\{u_n\}$ being a bounded sequence in the Hilbert space $H_*^1(M)$, we can find a weakly convergent subsequence $\{u_{1,n}\}$ in $H_*^1(M)$ with weak limit $v$ in $H_*^1(M)$. This subsequence being bounded in $H_*^1(M)$, its image $\{u_{1,n}\}$ in $L^2(M)$ is relatively compact and hence one can find a subsequence $\{v_n = u_{2,n}\}$ which converges weakly to $v$ in $H_*^1(M)$ and <u>strongly</u> to an element $u$ in $L^2(M)$. Since $\|v_n\|_o = 1$, we have $\|u\|_o = 1$. Since strong convergence implies weak convergence, $\{v_n\}$ converges weakly to $u$ in $L^2(M)$. The inclusion $H_*^1(M) \to L^2(M)$ being continuous, the $H_*^1(M)$ weak convergence of $\{v_n\}$ to $v$ implies the $L^2(M)$-weak convergence of $\{v_n\}$ to $v$ (viewed as an element of $L^2(M)$) and hence $u = v$.

Finally we have proved that $\{v_n\}$ converges $H_*^1(M)$-weakly and $L^2(M)$-strongly to an element $v$ in $H_*^1(M)$ such that $\|v\|_o = 1$.

From Cauchy-Schwarz inequality, we deduce that for any $f$ in $H_*^1(M)$

$$(v_n|f)_1^2 \le \|v_n\|_1^2 \; \|f\|_1^2 \le (\mu_1^* + 1 + \tfrac{1}{n})\|f\|_1^2.$$

It follows that

$$(v|f)_1^2 \le (\mu_1^*+1)\|f\|_1^2, \quad \text{and taking} \quad f = v,\ R(v) \le \mu_1^*.$$

Since $R(v) \ge \mu_1^*$, by definition of $\mu_1^*$, we conclude that the infimum $\mu_1^*$ of $R(u)$ on $H_*^1$ is achieved.

Let $E_1$ be the set of all elements $v$ in $H_*^1$ such that $v = 0$ or $v \ne 0$ and $R(v) = \mu_1^*$. Let $v \in E_1$. For any $u$ in

$H_*^1(M)$ and $t$ small enough in $\mathbb{R}$, we have $R(v+tu) \geq R(v) = \mu_1^*$. Writing that the derivative at $t = 0$ of the function $t \to R(v+tu)$ is zero, we have the following characterization of $E_1$

$$(22) \qquad v \in E_1 \Leftrightarrow \quad \forall\, u \in H_*^1, \ (u|v)_1 = (\mu_1^*+1)(u|v)_o.$$

From this characterization one can conclude that $E_1$ is a vector space. From the fact that the $\|.\|_1$-norm and $\|.\|_o$-norm are proportional on $E_1$, we conclude (Theorem 14) that the unit-$\|.\|_o$-ball of $E_1$ is compact and hence that $E_1$ is finite dimensional (for the elementary functional analysis we used above see [BZ]).

23. Summary. The infimum

$$\mu_1^* = \inf\{R(u) \mid u \in H_*^1(M),\ u \neq 0\}$$

is achieved on a finite dimensional subspace $E_1$ of $H_*^1(M)$ which is characterized by (22).

Given $u$ in $H_*^1(M)$, we also denote by $\nabla u$ the gradient of $u$ in the sense of distributions (in a local coordinate system $\nabla u = \sum_{i=1}^{n} \left( \sum_{j=1}^{n} g^{ij} \frac{\partial u}{\partial x_j} \right) \frac{\partial}{\partial x_i}$ where $\frac{\partial u}{\partial x_j}$ are derivatives in the sense of distributions).

The fact that $u$ belongs to $H_*^1(M)$ means that $|\nabla u|$ belongs to $L^2(M)$. Formula (22) can be written as follows: for any $u$ in $E_1$ and any $f$ in $H_*^1(M)$,

$$\int_M \langle \nabla u | \nabla f \rangle v_g = \mu_1^* \int_M uf\ v_g$$

which we can state as (see I.43-45)

"any element $u$ in $E_1$ is a weak solution of the eigenvalue problem (16*)"

(The boundary conditions are taken into account through $H_*^1(M)$). The classical regularity theory of weak solutions of elliptic problems ([G-T] or [TR2]) shows that $E_1$ is in fact contained in $C^\infty(M)$. Green's Theorem 10 finally shows that $u$ is in fact a classical solution of the eigenvalue Problem (16*): if $u$ is in $H^1_D \cap C^\infty$, then $u = 0$ on $\partial M$ and hence $u$ satisfies (16D); if $u$ is in $H^1_N \cap C^\infty$, then $\Delta u = \lambda u$ in $\mathring{M}$ (take $f$ in $C^\infty_o(M)$) and $\int_{\partial M} (\nu.u)f\, da_g = 0$ which implies that $\nu.u = 0$ on $\partial M$ (take $f$ in $C^\infty(M)$: compare with nº I.9 ff).

24. So far we have proved the existence of the first eigenvalue and its finite dimensional eigenspace. Let us denote by $L_1$ (resp. $H_1$) the subspace of $L^2(M)$ (resp. $H^1_*(M)$) which is orthogonal to $E_1$. Formula (22) shows that $H_1 \subset L_1$. These spaces are closed in $L^2(M)$ and $H^1_*(M)$ respectively and the inclusion $H_1 \subset L_1$ is compact.

We now define

$$\mu^*_2 = \inf\{R(u) \mid u \in H_1,\ u \neq 0\}.$$

Following the same arguments as those used above, we can prove that $\mu^*_2$ is indeed achieved on a finite dimensional subspace $E_2$ of $H_1$ which is characterized by

$$u \in E_2 \Leftrightarrow \text{for any } v \in H_1,\ \ (u|v)_1 = (\mu^*_2+1)(u|v)_o.$$

Noticing that the right-hand side equality holds trivially for $v$ in $E_1$, we deduce that $E_2$ is characterized by a formula analogous to (22) (change $\mu^*_1$ to $\mu^*_2$ in (22)). It is also clear that $\mu^*_2 > \mu^*_1$. We can construct an increasing sequence of non-negative real numbers

$$\mu^*_1 < \mu^*_2 < \ldots$$

and a sequence of associated finite dimensional subspaces of $H_*^1(M)$ which are mutually orthogonal (in $H_*^1(M)$ and in $L_*^2(M)$)

$$E_1, E_2, \ldots$$

Due to elliptic regularity theory, the functions in the $E_i$'s are $C^\infty$ and satisfy the eigenvalue Problem (16*). Notice that these sequences are infinite because $H_*^1(M)$ is infinite dimensional and the $E_i$'s are finite dimensional.

The sequence $\{\mu_i^*\}$ either increases to infinity or is bounded. If it were bounded by some number, we would have an infinite sequence $\{\phi_i\}$ of orthonormal functions in $L^2(M)$ (take $L^2$-orthonormal bases in the $E_j$'s) satisfying $R(\phi_i) \leq \mu$ and hence $\|\phi_i\|_1 \leq \mu+1$, in $H_*^1(M)$.
This is not possible because the inclusion $H^1(M) \to L^2(M)$ is compact.

Let $E$ denote the closure in $H_*^1(M)$ of the vector-space spanned by the vectors in the $E_j$'s. Assume $E \neq H_*^1(M)$. We can then find a function $u$ in $H_*^1(M)$, orthogonal to all the $E_j$'s in $H_*^1(M)$ or equivalently in $L^2(M)$ (because of (22)). It follows that $R(u) \geq \mu_i^*$ and $(u|u)_1 \geq (\mu_i^*+1)(u|u)_o$ for all $i$, which is impossible because $\mu_i^*$ tends to infinity. It follows that $\oplus E_i$ is dense in $H_*^1(M)$ and hence in $L^2(M)$. The other assertions in Theorem 18 follow from elliptic regularity results.

For more details see [SW] Chap. III.

25. Given $f$ in $H_*^1(M)$, one can write

$$R(f) = \sum_{i=1}^{\infty} \lambda_i a_i^2 \Big/ \sum_{i=1}^{\infty} a_i^2$$

where $a_i = (f|\phi_i)_o$, $f \neq 0$ (see nº 19).

This expression of $R(f)$ justifies a posteriori the second proof we sketched for Theorem 18; it also proves the following characterization of the eigenvalues and eigenfunctions of Problem (16*).

26. Variational Characterization I. (Notations as in nº 19)

The $k^{th}$ eigenvalue (with multiplicities) $\lambda_k$ is characterized by

$$\lambda_k = \inf\{R(u) \mid u \neq 0,\ u\ L^2\text{-orthogonal to } \phi_1,\dots,\phi_{k-1}\}$$

where $u$ is taken in $H^1_*(M)$ or in $C^\infty_*(M)$. Furthermore, if $u$ in $H^1_*(M)$ is $L^2$-orthogonal to $\phi_1,\dots,\phi_{k-1}$, and $R(u) = \lambda_k$ then $u$ is an eigenfunction of Problem (16*) associated with the eigenvalue $\lambda_k$.

Let us consider $\lambda_1 = \inf\{R(u) \mid u \text{ in } C^\infty_*(M),\ u \neq 0\}$. If we know enough functions $u$ on which we can calculate $R(u)$, then we know an upperbound for $\lambda_1$. This will turn out to be very important in the future. However, if we want upper bounds on $\lambda_2$ instead of $\lambda_1$, we have to know the eigenfunction $\phi_1$ and take $u$ $L^2$-orthogonal to $\phi_1$. Things are even more complicated with $\lambda_k$, $k \geq 3$. The following characterizations deal with these difficulties.

27. Variational Characterization II

The following variational characterization holds

$$\lambda_k = \sup_{M_{k-1}} \inf\{R(u) \mid u\ L^2\text{-orthogonal to } M_{k-1},\ u \neq 0\},$$

where $M_{k-1}$ runs through $(k-1)$-dimensional subspaces of $H^1_*(M)$ or $C^\infty_*(M)$.

Proof. Let $\Lambda(M_{k-1}) = \inf\{R(u) \mid u\ L^2\text{-orthogonal to } M_{k-1},\ u \neq 0\}$. Take $M^o_{k-1} = [\phi_1,\ldots,\phi_{k-1}]$ the vector-space spanned by the eigengunctions $\phi_1,\ldots,\phi_{k-1}$. Then $\Lambda(M^o_{k-1}) = \lambda_k$ according to the first variational characterization. This implies that

$$\sup_{M_{k-1}} \Lambda(M_{k-1}) \geq \lambda_k.$$

It is easy to show (by an argument on dimensions) that given a subspace $M_{k-1}$, one can find an element $v$ in $M^o_k = [\phi_1,\ldots,\phi_k]$ such that $v \neq 0$ and $v$ $L^2$-orthogonal to $M_{k-1}$. For such a $v$ one has $R(v) \leq \lambda_k$ and hence $\Lambda(M_{k-1}) \leq \lambda_k$. This implies that $\sup_{M_{k-1}} \Lambda(M_{k-1}) \leq \lambda_k$ ■

28. Variational Characterization III. (Notations as in nº 19)

The $k^{th}$ eigenvalue (with multiplicities) $\lambda_k$ is characterized by

$$\lambda_k = \inf_{L_k} \sup\{R(u) \mid u \text{ in } L_k,\ u \neq 0\}$$

where $L_k$ runs through k-dimensional subspaces of $H^1_*(M)$ or $C^\infty_*(M)$.

Proof. Taking $L_k = [\phi_1,\ldots,\phi_k]$ the vector-space spanned by the eigenfunctions $\phi_1,\ldots,\phi_k$, we find that

$$\Lambda_k = \inf_{L_k} \sup\{R(u) \mid u \text{ in } L_k,\ u \neq 0\} \text{ satisfies } \Lambda_k \leq \lambda_k.$$

Let $L_k$ be a k-dimensional subspace in $H^1_*(M)$ or $C^\infty_*(M)$. Then there exists an element $u$ in $L_k$ such that $u$ is orthogonal to $\phi_1,\ldots,\phi_{k-1}$. It follows from nº 25 that $R(u) = \sum_{j=k}^{\infty} \lambda_j a_j^2 / \sum_{j=k}^{\infty} a_j^2$ where $a_j = (u|\phi_j)_o$ and hence that $R(u) \geq \lambda_k$. We then deduce that $\Lambda_k \geq \lambda_k$ ■

For other characterizations see [BE] Chap. III.

The following proposition turns out to be useful when one wants to determine explicitely the eigenvalues and eigenfunctions of one of the Problems (16*).

29. Proposition

Let $(M,g)$ be a Riemannian manifold. Let $\{V_i\}_{i=1}^{\infty}$ be a sequence of non-trivial subspaces of $D_*(M)$ (see nº 20) with the following properties

(i) For all $i \geq 1$, there exists a real number $\mu_i$ such that, for all $f$ in $V_i$, $\Delta f = \mu_i f$;

(ii) The sum $\sum_{i=1}^{\infty} V_i$ (finite linear combinations of elements in the $V_i$'s) is dense in $L^2(M,v_g)$ for $\|\cdot\|_o$.

Then the sequence $\{\mu_i\}$ is the sequence of eigenvalues of Problem (16*), up to increasing rearrangement, and the $V_i$'s are the associated eigenspaces.

Proof. Exercise 31(2).

One can give an analogous statement at the level of Rayleigh quotient.

30. Proposition

Let $(M,g)$ be a Riemannian manifold and let $\{V_i\}_{i=1}^{\infty}$ be a sequence of non-trivial subspaces of $H_*^1(M)$ with the following properties

(i) For all $i \geq 1$ there exists a real number $\mu_i$ such that for all $u$ in $V_i$ and all $v$ in $H_*^1(M)$

$$(u|v)_1 = (\mu_i+1)(u|v)_o,$$

(ii) $\sum_{i=1}^{\infty} V_i$ is dense in $L^2(M)$ for $\|.\|_o$.

Then the sequence $\{\mu_i\}$ is the sequence of eigenvalues of the Rayleigh quotient in $H_*^1(M)$, up to increasing rearrangement, and the $V_i$'s are the associated eigenspaces.

Proof. Take $(\mu_i^*, E_i)$ as in the proof of Theorem 18.

For $u$ in $V_i$ and $v$ in $E_j$ we can write

$$(u|v)_1 = (\mu_i+1)(u|v)_o, \quad \text{and (formula (22))}$$

$$(u|v)_1 = (\mu_j^*+1)(u|v)_o.$$

We then conclude that either $(u|v)_o = 0$ or $\mu_i = \mu_j^*$. This fact and Assertion (ii) show that the sequences $\{\mu_i\}$ and $\{\mu_i^*\}$ are equal. The characterization of the $E_i$'s (Formula (22)) shows that (up to an increasing rearrangement on the $\mu_i$'s) $V_i \subset E_i$. Assertion (ii) shows that $V_i = E_i$, because $V_i$ is orthogonal to $E_j$, $j \neq i$. ■

## 31. Exercises

(a) Prove Proposition 29;

(b) Let $(M,g)$ (resp. $(N,h)$) be a Riemannian manifold without boundary, with eigenvalues $\bar{\lambda}_i^M$ (resp. $\bar{\lambda}_i^N$) and eigenspaces $E_i^M$ (resp. $E_i^N$).

Find the eigenvalues and eigenspaces of the product Riemannian manifold $(M \times N, \; g \times h)$;

(c) Let $(N,h) \xrightarrow{p} (M,g)$ be a finite (Riemannian) covering i.e. $N \xrightarrow{p} M$ is a finite covering of manifolds without boundaries and $p^*g = h$.

Describe the eigenfunctions of $(M,g)$ in terms of the eigenfunctions of $(N,h)$ (see [B-G-M] Chap. III Prop. A.II.5 p.145);

(d) Let $\Omega$ be a smooth bounded domain in $(\mathbb{R}^n, \text{can})$.

Show that for all $i$ (notations in nº 19)

$$\lambda_i(\Omega, N) \leq \lambda_i(\Omega, D)$$

(Hint: use nº 26-28);

(e) Let $\Omega_1 \subset \Omega_2$ be two smooth bounded domains in $(\mathbb{R}^n, \text{can})$. Show that for all $i$ (Notations in nº 19)

$$\lambda_i(\Omega_1, D) \geq \lambda_i(\Omega_2, D)$$

(Hint: use nº 26-28);

(f) Let $\Omega_{a,b}$ be a rectangle with sides a and b in $(\mathbb{R}^2, \text{can})$. Find the eigenvalues and eigenfunctions of Problems (16D) and (16N) in $\Omega_{a,b}$ (Hint: use separation of variables and Proposition 29). When are all the eigenvalues Problem (16D) or Problem (16N) in $\Omega_{a,b}$ simple?

(g) Give upper and lower bounds for $\lambda_1(B(0,1), D)$ where $B(0,1)$ is the unit ball in $(\mathbb{R}^n, \text{can})$ (Hint: use nº 26-28 and generalize Exercise (f));

## C. EIGENVALUE PROBLEMS ON RIEMANNIAN MANIFOLDS II.

32. Much research effort has been devoted to eigenvalue problems since the $18^{\text{th}}$ century. These problems arise from (linear) mathematical models for questions in mathematical physics: acoustics, elasticity, plasma physics, spectroscopy, wave guides... These problems can be roughly divided into two types: direct problems and inverse problems.

In <u>a direct problem</u>, one seeks information about the eigenvalues and the eigenfunctions of the Laplacian $\Delta^g$ in terms of geometrical data. It turns out that it is usually not possible to determine explicitely the eigenvalues or the eigenfunctions (except when symmetries allow to reduce the original problem to one-dimensional eigenvalue problems: see Exercise 30(f) and [CL] Chap.II). Important progress have been made in the field of high-speed computers which allow reliable numerical computations of eigenvalues. Although these numerical methods are now used extensively, they do not discard theoretical investigations on the eigenvalues and eigenfunctions (see [K-S]).

A very important theoretical problem consists in finding bounds on the eigenvalues. Due to the variational characterizations of eigenvalues (nº 26-28) it is easier to obtain upper bounds than lower bounds. It turns out that lower bounds are more interesting from both the mathematical and physical points of view. For example, lower bounds determine safety limits of some mechanical systems in order to avoid buckling: rods, plates, beams.

The most powerful methods which have been developped in order to obtain bounds on the eigenvalues are the <u>isoperimetric methods</u>. These methods owe very much to the works of G. Polya and G. Szegö in the 40s (see [P-S] or [PE]); we will study them in Chapters IV-V of these notes.

In <u>an inverse problem</u>, one assumes that one or several eigenvalues of the Laplacian $\Delta^g$ are known and one seeks information on the metric $g$: curvature, form (i.e. topology) of the manifold.... Let us quote Sir A. Schuster (1882) who created the word <u>spectroscopy</u>: "to find out the different tunes sent out by a vibrating system is a problem which may or may not be solvable in certains cases, but it would baffle the most skillful mathematician to solve the inverse problem and to find out the shape of a bell by means of the sounds which it is capable of sending out. And this is the problem which ultimately spectroscopy hopes to solve in the case of light. In the meantime we must welcome with delight even the smallest step in the desired direction" (quoted in [G-S] Introduction p.8).

We will not deal with inverse problems in these notes but for a brief survey (Chap. VII).

However, we shall be interested in an <u>inverse geometric problem</u>; one important question in Riemannian geometry is to determine the <u>global influence</u> on the manifold of <u>(local) estimates</u> on the curvature. Theorem II.29 (Myers theorem) gives a partial answer to this question in dimension $n$; the Gauss-Bonnet theorem also gives a partial answer in the two-dimensional case ([HF] Theorem III p.113). Chapter VI is devoted to an analytic approach to the above question. In Chapter V-VI, we shall show how local estimates on the curvature (after scaling the metric appropriately) imply bounds on geometric invariants such as the Betti numbers of the manifold.

Let us now give two examples, one example of an inverse problem (H. Weyl's asymptotic formula) and one example of a direct problem (S.Y. Cheng's upper bounds on the eigenvalues).

33. Let $\Omega$ be a smooth bounded domain in $(\mathbb{R}^n, \text{can})$ i.e. with the usual Euclidean structure and Laplacian. We consider the Dirichlet eigenvalue problem (16D) in $\Omega$

$$\begin{cases} \Delta u = \lambda u & \text{in } \Omega \\ u = 0 & \text{on } \partial\Omega. \end{cases}$$

Let us consider a grid with size a in $\mathbb{R}^n$ i.e. the pattern of cubes in $\mathbb{R}^n$ made by a lattice $(a\mathbb{Z})^n$ centered at an interior point in $\Omega$: see Figure 3.

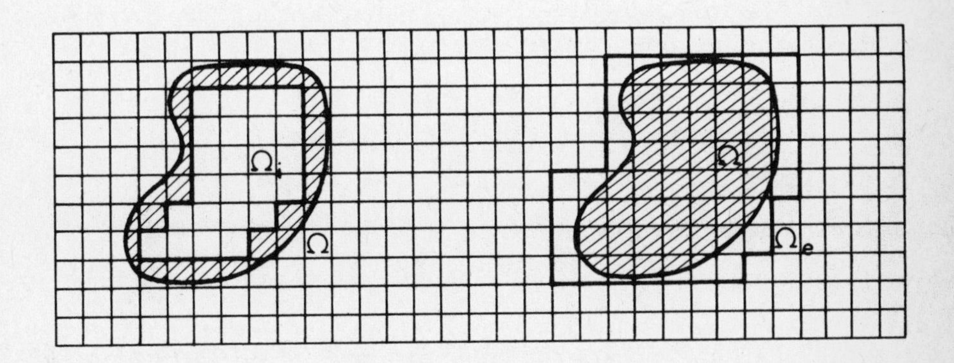


Fig. 3

Let us call $\mathcal{C}_i$ (resp. $\mathcal{C}_e$) the collection of all cubes of the grid which are contained in $\Omega$ (resp. which contain a point lying in $\Omega$).

Let $\mathcal{E}_i$ (resp. $\mathcal{E}_e$) be the collection of all eigenvalues (with multiplicities) of all the cubes in $\mathcal{C}_i$ (resp. $\mathcal{C}_e$) with Dirichlet (resp. Neumann) boundary condition: an eigenvalue which appears for two different cubes should be counted twice. Arrange the sets $\mathcal{E}_i$ and $\mathcal{E}_e$ in increasing sequences $\{\mu_j^{(i)}\}_{j=1}^{\infty}$ and $\{\mu_j^{(e)}\}_{j=1}^{\infty}$.

Denote by $\{\lambda_j\}_{j=1}^{\infty}$ the increasing sequence of the eigenvalues of the Dirichlet problem in $\Omega$.

34. Proposition. For $j \geq 1$ the following inequalities hold

$$\mu_j^{(e)} \leq \lambda_j \leq \mu_j^{(i)}.$$

Proof. Denote the generic cube of the grid by $C$. Let

$$L_i = \bigoplus_{C \in \mathcal{C}_i} L^2(C), \quad L_e = \bigoplus_{C \in \mathcal{C}_e} L^2(C)$$

$$H_i = \bigoplus_{C \in \mathcal{C}_i} H_o^1(C), \quad H_e = \bigoplus_{C \in \mathcal{C}_e} H^1(C)$$

We can view the (open) cubes as disjoint manifolds. We then have two manifolds $M_i = \coprod_{C \in \mathcal{C}_i} C$, $M_e = \coprod_{C \in \mathcal{C}_e} C$ (disjoint unions).

In that case, we also have

$$L_i = L^2(M_i), \ H_i = H_o^1(M_i), \ L_e = L^2(M_e), \ H_e = H^1(M_e).$$

Proposition 31 shows that $\{\mu_j^{(i)}\}$ (resp. $\{\mu_j^{(e)}\}$) is the sequence of the eigenvalues of the Rayleigh quotient on $H_o^1(M_i)$ (resp. $H^1(M_e)$). In particular, we have the following variational characterization

$$\mu_k^{(i)} = \inf_{L_k} \sup\{R(u) \mid u \text{ in } L_k,\ u \neq 0\}$$ (resp. $\mu_k^{(e)}$), where $L_k$ runs through the k-dimensional subspaces of $H_i$ (resp. $H_e$).

Let $\Omega_i = \bigcup_{C \in \mathcal{C}_i} \bar{C}$, $\Omega_e = \bigcup_{C \in \mathcal{C}_e} \bar{C}$. It is clear that $\Omega_i \subset \Omega \subset \Omega_e$.

Since $H_i \subset H_o^1(\Omega_i) \subset H_o^1(\Omega)$ we have

$$\mu_k^{(i)} \geq \lambda_k(\Omega_i, \text{Dirichlet}) \geq \lambda_k(\Omega, \text{Dirichlet}).$$

Since $H_o^1(\Omega) \subset H^1(\Omega_e) \subset H_e$ we have

$$\lambda_k = \lambda_k(\Omega, \text{Dirichlet}) \geq \lambda_k(\Omega_e, \text{Neumann}) \geq \mu_k^{(e)}.$$

Finally, we obtain $\mu_k^{(e)} \leq \lambda_k \leq \mu_k^{(i)}$ ■

35. Let $N(\lambda) = \text{Card}\{j \mid \lambda_j \leq \lambda\}$,

$$N^{(e)}(\lambda) = \text{Card}\{j \mid \mu_j^{(e)} \leq \lambda\},$$

$$N^{(e)}(\lambda) = \text{Card}\{j \mid \mu_j^{(i)} \leq \lambda\}.$$

<u>Lemma</u>. <u>When</u> $\lambda$ <u>goes to infinity the following limits hold</u>

$$\lim \lambda^{-n/2} N^{(e)}(\lambda) = c(n) \text{ Vol}(\Omega_e)$$

$$\lim \lambda^{-n/2} N^{(i)}(\lambda) = c(n) \text{ Vol}(\Omega_i)$$

<u>where</u> $c(n) = (2\pi)^{-n}.\text{Vol}$ unit ball in $\mathbb{R}^n$.

<u>Proof</u>. The eigenvalues of a cube with side a are of the form

$$\frac{\pi^2}{a^2} \sum_{\ell=1}^{n} k_\ell^2 \quad \text{where}$$

$$k_\ell \in \mathbb{N}(\text{Neumann}) \quad \text{or} \quad k_\ell \in \mathbb{N} \setminus \{0\} \quad (\text{Dirichlet})$$

Hint: use separation of variables and Theorem 30.

Counting the eigenvalues less than $\lambda$ amounts to counting the points with positive integer coordinates inside the ball of radius $\frac{a}{\pi}\sqrt{\lambda}$. As $\lambda$ goes to infinity the equivalent for the number of such points is $(2\pi)^{-n}$(Vol unit ball in $\mathbb{R}^n$) $a^n \lambda^{n/2}$ ($2^{-n}$ appears because we only consider points with positive coordinates). We can interpret $a^n$ as the volume of a generic cube in the grid. The Lemma follows from the definitions of $\mu_k^{(i)}$ and $\mu_k^{(e)}$ ■

36. Theorem (Weyl's asymptotic formula)

Let $\Omega$ be a smooth bounded set in $(\mathbb{R}^n, \text{can})$.

Let $\{\lambda_k\}_{k=1}^{\infty}$ denote the sequence of eigenvalues of the Dirichlet eigenvalue problem for the Laplacian in $\Omega$. Let $N(\lambda) = \text{Card}\{j \mid \lambda_j \leq \lambda\}$. The asymptotic behavior of $N(\lambda)$ is given by

$N(\lambda) \sim c(n)\text{Vol}(\Omega)\lambda^{n/2}$ as $\lambda \to \infty$ ($c(n) = (2\pi)^{-n}$Vol unit ball in $\mathbb{R}^n$).

Proof. Take a grid as above and notice that one can take $\text{Vol}(\Omega_i)$ and $\text{Vol}(\Omega_e)$ very close to $\text{Vol}(\Omega)$ by taking a small. Compare with [C-H] Chap. VI.4 and Chap. VII.14 [R-S], §XIII.15■

37. Remarks

(i) Weyl's formula also holds for any of the eigenvalue Problems (16): see nº VII.11;

(ii) Weyl's formula shows that if we know all the eigenvalues of $\Omega$, say for the Dirichlet eigenvalue problem, then we know the dimension $n$ of $\Omega$ and its volume $\text{Vol}(\Omega)$. This is an example of an answer to an inverse problem: the knowledge of the Dirichlet-spectrum of $\Omega$ gives both the dimension and the volume of $\Omega$.

38. Corollary. For $\Omega$ as in Theorem 36 we have

$$\lambda_j(\Omega) \underset{j\to\infty}{\sim} c(n)^{-2/n}\left(\frac{j}{\text{Vol}(\Omega)}\right)^{2/n} .$$

Our next result is an example of an answer to a *direct problem*.

39. *Theorem. Let* $(M,g)$ *be an* $n$*-dimensional Riemannian manifold without boundary, whose Ricci curvature is bounded from below by* $(n-1)k$ $(k \text{ in } \mathbb{R})$:

$$\mathrm{Ricci}(M,g) \geq (n-1)kg.$$

*Let* $r$ *be less than the injectivity radius of* $(M,g)$.

*For any* $x$ *in* $M$, *the following inequality holds*

$$\lambda_1(B(x,r),\mathrm{Dirichlet}) \leq \lambda_1(k,r)$$

*where* $\lambda_1(k,r)$ *is the first eigenvalue for the Dirichlet problem in a geodesic ball of radius* $r$ *in the space* $(\mathbb{S}^n_k,\mathrm{can})$ *with constant curvature* $k$ (*see* nº II.18).

40. *Comments*.

(1) The *injectivity radius* $\mathrm{Inj}(M,g)$ is the largest $r$ such that for all $x$ in $M$, $\exp_x$ is an embedding on the open ball of radius $r$ in $T_mM$. When $M$ is compact this number is (strictly) positive ([C-E] Chap. 5).

(2) It follows from Theorem 18 and Green's formula (Theorem 10) that for a Riemannian manifold with boundary $(N,h)$, the first eigenvalue $\lambda_1(N,h,D)$ of the Dirichlet eigenvalue problem is (strictly) positive. For the Closed or Neumann eigenvalue problems the first eigenvalue is $0$.

(3) All the geodesic balls of radius $r$ in $(\mathbb{S}^n_k,\mathrm{can})$ are isometric so that the definition of $\lambda_1(k,r)$ in the theorem makes sense.

(4) In fact, Theorem 39 also holds for radii larger than $\mathrm{Inj}(M,g)$: see S.Y. Cheng, [CG].

Proof. It can be shown ([CL], Chap. II.5) that the first eigenfunction of the Dirichlet eigenvalue problem in the ball $B(p,r)$ in $(S_k^n, \mathrm{can})$ can be written as $\phi_1 = \varphi(d_k(p,.))$ where $\varphi$ is a positive function and $d_k(p,.)$ is the Riemannian distance function to $p$ in $(S_k^n, \mathrm{can})$.

Let $\bar{\theta}(r,u) = \bar{\theta}(r)$ be the volume density in $(S_k^n, \mathrm{can})$ viewed through $\exp_p$ (see nº II.25). The function $\varphi$ satisfies

$$\varphi''(s) + \frac{\bar{\theta}'(r)}{\bar{\theta}(r)} \varphi'(s) + \lambda_1(k,r)\varphi(s) = 0;$$

$$\varphi(r) = 0;$$

$$\varphi \; C^\infty.$$

From this equation it follows that $\varphi$ is decreasing.

For $x$ in $M$, let $f(y) = \varphi(d(x,y))$ where $y \in B(x,r)$ and $d(x,y)$ is the Riemannian distance in $(M,g)$ (this procedure is called transplantation: see [BE], [P-S]).

From the first variational characterization of eigenvalues (nº 26) we can write

$$\lambda_1(B(x,r)) \leq \int_{B(x,r)} |df|^2 \Big/ \int_{B(x,r)} f^2 \quad \text{because } f$$

vanishes on the boundary of $B(x,r)$. Let $\theta(s,u)$ denote the volume density in $(M,g)$ viewed through $\exp_x$ (nº II.25).

Pulling back the above integrals to $T_xM$ we obtain

$$\int_{B(x,r)} |df|^2 = \int_{S^{n-1}} \int_0^r (\varphi'(s))^2 \theta(s,u) ds\, du,$$

$$\int_{B(x,r)} f^2 = \int_{S^{n-1}} \int_0^r \varphi^2(s)\theta(s,u)ds\, du.$$

Integration by parts gives

$$\int_{B(x,r)} |df|^2 = \int_{S^{n-1}} \int_0^r \varphi(s)\{-\varphi''(s) - \frac{\theta'}{\theta}(s,u)\varphi'(s)\}\theta(s,u)ds\, du.$$

41. Lemma. *With the above notations and under the assumption* Ricci $(M,g) \geq (n-1)kg$, *we have*

$$\frac{\partial}{\partial s}\{\theta(s,u)/\bar{\theta}(s)\} \leq 0$$

*for* $s$ *smaller than* $\mathrm{Inj}(M,g)$.

For a proof see [B-C] Chap. 11.10; this Lemma is the key-point in the proof of the Bishop-Gromov comparison theorem (nºII.30).

Now since $\varphi'(s) < 0$ we conclude that

$$\int_B |df|^2 \leq \int_{S^{n-1}} \int_0^r \varphi(s)\{-\varphi''(s) - \frac{\bar{\theta}'}{\bar{\theta}}(s)\varphi'(s)\}\theta(s,u)ds\, du =$$

$$= \lambda_1(k,r) \int_{S^{n-1}} \int_0^r \varphi^2(s)\theta(s,u)ds\, du \quad ■$$

42. Corollary. *Let* $(M,g)$ *be an* $n$-*dimensional Riemannian manifold without boundary, whose Ricci curvature is bounded from below by* $(n-1)k$ *i.e.* $\mathrm{Ricci}(M,g) \geq (n-1)kg$. *Let* $D$ *denote the diameter of* $(M,g)$ *and let* $\{\lambda_i(M)\}_{i=1}^{\infty}$ *denote the eigenvalues of the closed eigenvalue problem on* $(M,g)$ *counted with multiplicities.*

*The following inequalities holds*

$\lambda_m(M) \leq \lambda_1(k, D/2(m-1))$, *for* $m \geq 2$ (*recall that* $\lambda_1(M)=0$),

*where* $\lambda_1(k,r)$ *denotes the first eigenvalue for the Dirichlet* eigenvalue problem in a ball of radius $r$ in the space $(\$^n_k, can)$.

Proof. Take $x$ and $y$ in $(M,g)$ such that $d(x,y) = D$ and consider a shortest path (see II.8) from $x$ to $y$. One can find $x_1,\ldots,x_m$ on this path such that the open balls $B(x_i, D/2(m-1))$ are pair-wise disjoint.

Consider the vector-space in $H^1(M)$ spanned by the $m$ functions $f_1,\ldots,f_m$, where

$$f_i = \begin{cases} 1^{st} \text{ eigenfunction (for Dirichlet) in } B(x_i, D/2(m-1)), \\ 0 \quad \text{outside } B(x_i, D/2(m-1)). \end{cases}$$

This subspace has dimension $m$. By Theorem 39, for any $u$ in this subspace $R(u) \le \lambda_1(k, D/2(m-1))$.

Corollary 42 then follows from the third variational characterization of eigenvalues (nº 28) ■

43. Comments.

Corollary 42 says that the eigenvalues of $(M,g)$ can be estimated from above in terms of a lower bound on the Ricci curvature and an upper bound on the diameter. The estimate given in Corollary 42 is not satisfactory for several reasons.

For $k$ and $n$ fixed, we have $\lim_{r\to 0_+} r^2\lambda_1(k,r) = c(k,n)$ (this follows from the fact that a Riemannian manifold is asymptotically Euclidean). It then follows from Corollary 42 that $\lim_{m\infty} D^2 m^{-2}\lambda_m(M) \le c_1(k,n)$. Recall that Weyl's asymptotic formula reads $\lim_{m\infty} m^{-2/n}\,\mathrm{Vol}(M)^{2/n}\lambda_m(M) = c(n)$, and notice that both $\mathrm{Diam}^2(M,g)\lambda_m(M,g)$ and $\mathrm{Vol}(M,g)^{2/n}\lambda_m(M,g)$ are Riemannian invariants of weight 0 (see nº II.6).

It is easy to explain why Corollary 42 does not give an estimate compatible with Weyl's asymptotic formula. We took a shortest path from $x$ to $y$, so that we acted as if $(M,g)$ were one-dimensional and we therefore found $m^{-2/1}$ instead of $m^{-2/n}$ and its "1-dimensional volume" $\mathrm{Diam}(M) = D$ instead of $\mathrm{Vol}(M,g)$, its n-dimensional volume.

In order to get closer to Weyl's formula, we have to fill the Riemannian manifold with balls. The following argument is due to M. Gromov ([GV1]). For a given $\varepsilon > 0$, let $\{x_i\}_{i=1}^{N(\varepsilon)}$ denote a maximal set of points in $M$ such that the balls $B(x_i,\varepsilon)$ are pair-wise disjoint and $\bigcup_{i=1}^{N} B(x_i,2\varepsilon) = M$. Let $b_k(r)$ denote the volume of a geodesic ball with radius $r$ in $\$_k^n$.

Applying Bishop's comparison Theorem (II.30(i)), we can write

$$\mathrm{Vol}(M,g) \leq \sum_{i=1}^{N(\varepsilon)} \mathrm{Vol}\, B(x_i,2\varepsilon) \leq N(\varepsilon)\, b_k(2\varepsilon).$$

The 3rd variational characterization of eigenvalues (nº 28) and Theorem 39 give

$$(44) \qquad \lambda_{N(\varepsilon)}(M,g) \leq \lambda_1(k,\varepsilon).$$

For $k$ and $n$ fixed, we have

$$\lim_{\varepsilon\to 0} \varepsilon^2 \lambda_1(k,\varepsilon) = c_1(k,n)$$

$$\lim_{\varepsilon\to 0} b_k(2\varepsilon)\varepsilon^{-n} = c_2(k,n)$$

$$\lim_{\varepsilon\to 0} N(\varepsilon)\varepsilon^n \geq \frac{\mathrm{Vol}(M,g)}{c_2(k,n)}$$

and finally we obtain the estimate

$$\lim_{m\infty} \lambda_m(M,g)\, \mathrm{Vol}(M,g)^{2/n} m^{-2/n} \leq c_3(k,n)$$

which is very close to Weyl's estimate.

In fact, (44) shows that one can estimate the eigenvalues $\lambda_m(M,g)$ from above in terms of a lower bound on the Ricci curvature and the volume.

In [GV1], M. Gromov also gave lower bounds for the eigenvalues $\lambda_m(M,g)$, in terms of Ricci and Diameter.

More precisely, let $r_{min} = \inf\{Ricci(u,u) \mid u \in UM\}$. Assume that $r_{min} Diam^2(M,g) \geq (n-1)\alpha$. Then there exists a constant $C(\alpha,n)$ such that

$$(45) \qquad \lambda_m(M,g)\, Diam(M,g)^2 \geq C(\alpha,n) m^{2/n}$$

for all $m \geq 2$.

We shall prove such an estimate in Chapter VI (our constant will be better than Gromov's and our method quite different). In view of Weyl's asymptotic formula, one can ask whether one can replace $Diam(M,g)^2$ by $Vol(M,g)^{2/n}$ in (45), as we did above for the upper bounds. In V.32, we will give a counter-example showing that (45) is best possible qualitatively: a general lower bound on $\lambda_m(M,g)$ must depend on a lower bound on the Ricci curvature and an upper bound on the diameter.

46. <u>Remarks</u>.

(1) Theorem 39, Corollary 42 and (44) give partial answers to a <u>direct problem</u> (find information on the eigenvalues in terms of geometric data).

(2) As the variational characterizations show, to a stronger stress correspond larger eigenvalues of given rank. This should be kept in mind together with our motivations from mathematical physics in Chapter I.

Further references for Chapter III: [AN], [B-J-S], [B-G-M], [BN], [BZ], [C-H], [CL], [ES], [WS].

# CHAPTER IV

## ISOPERIMETRIC METHODS

In this chapter, we give the basic ideas concerning the isoperimetric methods together with some direct applications.

### A. MOTIVATIONS: THE FABER-KRAHN INEQUALITY.

Given $\Omega$, a smooth bounded domain in $\mathbb{R}^n$, we denote by $\lambda_1(\Omega)$ the first eigenvalue for the Dirichlet eigenvalue problem in $\Omega$; we denote by $\Omega^*$ the Euclidean ball centered at 0 in $\mathbb{R}^n$, whose volume is equal to $\mathrm{Vol}(\Omega)$. The following inequality holds

(1) $\lambda_1(\Omega) \geq \lambda_1(\Omega^*)$ (Faber-Krahn's inequality).

This inequality was stated in dimension 2 by Lord Rayleigh in his treatise on the theory of sound ([RH], Section 210; this is still a very stimulating reading). The proof of inequality (1) was given independently by C. Faber and E. Krahn in the 1920s.

2. Sketch of the proof of the Faber-Krahn inequality.

3. Lemma. *Let f be any eigenfunction associated with the first eigenvalue $\lambda_1$ of the Dirichlet problem in $\Omega$. Then f is (strictly) positive or (strictly) negative in the open set $\Omega$.*

Proof. Since $f \in H_o^1(\Omega)$, we have $|f| \in H_o^1(\Omega)$ and $R(f)=R(|f|)=\lambda_1$ ([G-T] §7.4). It follows that $|f|$ also is an eigenfunction associated with $\lambda_1$ (III.26) and hence $|f| \in C^2(\Omega) \cap C^0(\bar{\Omega})$ by elliptic regularity theory ([G-T] Chap. 8). Now, $\Delta(|f|) = \lambda_1 |f| \geq 0$ (recall our sign convention on $\Delta$) and it follows that $\inf_{\bar{\Omega}} |f|$ is achieved only on $\partial\Omega$, unless $f \equiv 0$ (maximum principle, [G-T] Theorem 3.5 p.34). Finally, we have $|f| > 0$ in $\Omega$ and hence $f > 0$ or $f < 0$ ■

Another reference for the maximum principle is M. Spivak [SK], Vol. V, Addendem 2 to Chapter 10 p. 181ff. In order to prove Lemma 3, one can also use a local argument (Taylor's expansion near a point where $f$ vanishes in $\Omega$) and the unique continuation property ([AZ]) ■

4. Corollary. The first eigenvalue $\lambda_1$ for the Dirichlet problem in $\Omega$ is simple.

Proof. Assume not. Take $f_1, f_2$ two orthogonal eigenfunctions. They can also be assumed to be positive in $\Omega$ by Lemma 3, and hence $\int_\Omega f_1 f_2 > 0$, a contradiction ■

Let $f$ be the first eigenfunction in $\Omega$ associated with $\lambda_1(\Omega)$.

The main idea in the proof of inequality (1) is the idea of symmetrization.

We consider the sets $\Omega_t = \{x \in \Omega \mid f(x) > t\}$ and we symmetrize them by considering the Euclidean balls $\Omega_t^*$ in $\mathbb{R}^n$, with center 0, satisfying $\mathrm{Vol}(\Omega_t) = \mathrm{Vol}(\Omega_t^*)$.

Equivalently, we symmetrize the graph $F$ of $f$ above $\Omega$ into a set $F^*$ which is invariant under rotations about the axis $D^*$ (Fig. 4).

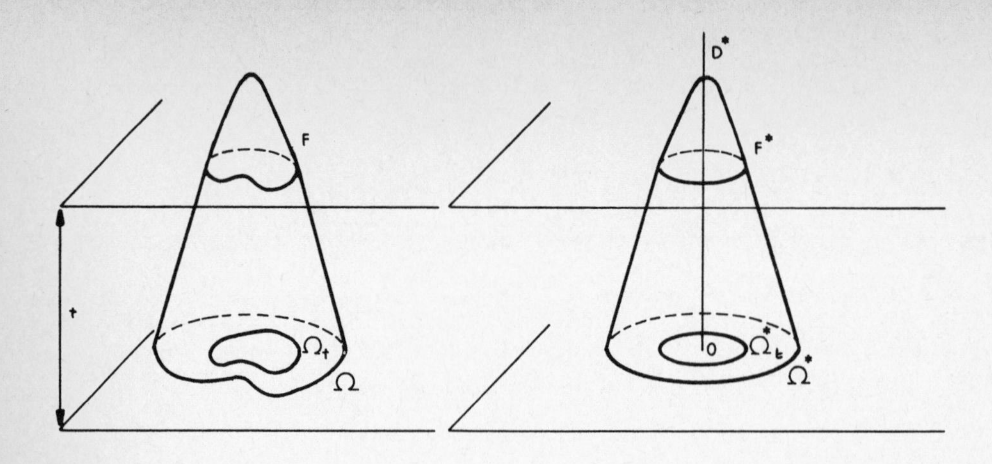


Fig. 4

We define a function $f^*$ on $\Omega^*$ by the following properties

(i) the graph of $f^*$ is $F^*$, or equivalently,

(ii) $\begin{cases} f^* \text{ is a radial decreasing function} \\ \text{and} \\ f^* \text{ takes the value } t \text{ on } \partial\Omega_t^*. \end{cases}$

This procedure is called the <u>symmetrization of the function</u> f. (we do not wish to go into formal definitions now: for more details see [BE] Chap. II, [MO] Chap. I, Appendix A or [KL]).

In order to estimate the Rayleigh quotient $R(f)$ of the function $f$, we introduce new coordinates on $\Omega$, by considering the level hypersurfaces and the lines of gradient of $f$. The following formula is known as the <u>co-area formula</u> (see [CL] Chap. IV, [BE] Lemma 2.5 p. 53, [B-M] Appendix A, [FR]).

5. Lemma.

For any continuous function $h$ on $\Omega$ one has

$$\int_{\Omega} h(x)dx = \int_0^{\sup f} \left(\int_{G(t)} h|df|^{-1}\, da_t\right)dt,$$

where $da_t$ is the volume element of the Riemannian metric induced by $\mathbb{R}^n$ on the hypersurface $G(t) = f^{-1}(t)$ (this makes sense for $t$ in the set $\mathcal{R}_f$ of regular values of $f$; the complement of $\mathcal{R}_f$ has measure zero by Sard's theorem).

If we now take $h = |df|^2$ and if we apply the co-area formula, we obtain $(m=\sup f)$

$$\int_{\Omega} |df|^2(x)dx = \int_0^m \left(\int_{G(t)} |df|\, da_t\right)dt.$$

Applying Cauchy-Schwarz inequality, we find

$$\int_{G(t)} |df|\, da_t \geq \left(\int_{G(t)} da_t\right)^2 \Big/ \int_{G(t)} |df|^{-1} da_t$$

(for $t$ in $\mathcal{R}_f$).

Now $\int_{G(t)} da_t$ is just the $(n-1)$-dimensional volume of $G(t) = f^{-1}(t)$, hence, by the classical isoperimetric inequality in $\mathbb{R}^n$ (see nº 8 below), $\left(\int_{G(t)} da_t\right)^2 \geq \mathrm{Vol}(\partial\Omega_t^*)^2$.

It also follows from the co-area formula that

$$-\int_{G(t)} |df|^{-1} da_t = \frac{d}{dt}\mathrm{Vol}(\Omega_t) = \frac{d}{dt}\mathrm{Vol}(\Omega_t^*).$$

If we now apply the same construction to the radial function $f^*$, (see [CL] Chap. IV or [B-M] Appendix B) we have

$$\int_{G^*(t)} |df^*| da_t^* = \left(\int_{G^*(t)} da_t^*\right)^2 / \int_{G^*(t)} |df^*|^{-1} da_t^*$$

($|df^*|$ is constant on $G^*(t) = f^{*-1}(t)$) and hence,

$\int_{G(t)} |df| da_t \geq \int_{G^*(t)} |df^*| da_t^*$. Integrating in $t$, this gives

$$\int_\Omega |df|^2(x)dx \geq \int_{\Omega^*} |df^*|^2(x)dx.$$

It follows easily from the co-area formula that

$$\int_\Omega f^2(x)dx = \int_{\Omega^*} f^{*2}(x)dx.$$

Finally we have proved that $R(f;\Omega) \geq R(f^*;\Omega^*)$, from which it follows that $\lambda_1(\Omega) \geq \lambda_1(\Omega^*)$ (First variational characterization of the eigenvalues: III.26).

6. Remarks

(i) One way of dealing with the difficulties arising from the co-area formula is to approximate functions in $H_o^1(\Omega)$ by "nice" Morse functions (this argument was introduced by Th. Aubin: see [B-M] Lemma 10 bis p. 519);

(ii) One can also use a more general form of the co-area formula: see [TI], [MO], [KL], [FR];

(iii) The Faber-Krahn inequality can be generalized to other situations: see [CL] Chap. IV, [B-M] and nº 22 below.

7. Comments.

The main ideas in the proof are the principle of symmetrization and the use of the classical isoperimetric inequality (both Cauchy-

Schwarz inequality and the co-area formula are technical details which are easily generalized to other situations). The classical <u>isoperimetric inequality</u> in $\mathbb{R}^n$ states that among all domains in $\mathbb{R}^n$, with given n-dimensional volume, the Euclidean balls have least boundary (n-1)-dimensional volume. Because dilations act on $\mathbb{R}^n$, this inequality can be written as follows: for any bounded domain $\Omega$ in $\mathbb{R}^n$,

$$(8) \qquad \mathrm{Vol}_{n-1}(\partial\Omega) \geq C^*(n)\, \mathrm{Vol}_n(\Omega)^{(n-1)/n},$$

where $C^*(n) = \mathrm{Vol}_{n-1}(S^{n-1})/\mathrm{Vol}_n(B^n)^{(n-1)/n}$,

$$B^n = \{x \text{ in } \mathbb{R}^n \mid |x| = 1\}, \quad S^{n-1} = \partial B^n.$$

(For a more general statement, for example when $\partial\Omega$ is very irregular, see [FR] p. 278).

Inequality (8) explains our choice of the symmetrization procedure, so that the principle of symmetrization and the isoperimetric inequality amount to the same idea. This idea can be generalized to analogous situations on the sphere $(S^n,\mathrm{can})$ or on the hyperbolic space $(H^n,\mathrm{can})$: <u>among domains of</u> $(\$^n_k,\mathrm{can})$ <u>with given volume, the geodesic balls have least boundary volume</u> (see [ON] for a survey on isoperimetric inequalities).

The point is that the model spaces $(\$^n_k,\mathrm{can})$ have many isometries and nice geodesic balls. We cannot expect anything like that on a generic Riemannian manifold. We will now explain how the symmetrization procedure can be extented to the general case.

## B. ISOPERIMETRIC INEQUALITIES AND SYMMETRIZATION

Although some of the ideas we will deal with can be generalized to other situations, we will from now on assume that

ALL RIEMANNIAN MANIFOLDS ARE COMPACT, CONNECTED, WITHOUT BOUNDARY.

9. An isoperimetric inequality on a Riemannian manifold $(M,g)$ (compact, without boundary) is an estimate from below of the volume of the boundary $\partial\Omega$ of a domain $\Omega$ in $M$, in terms of $\mathrm{Vol}(\Omega)$. If we take $\Omega$ to be the complement of a small ball in $M$, we see that it is more realistic to consider $\mathrm{Vol}(\Omega)/\mathrm{Vol}(M)$, the relative volume of $\Omega$ in $M$, instead of $\mathrm{Vol}(\Omega)$.

We define the isoperimetric function of $(M,g)$ as

$$h(\beta) = h(M,g;\beta) = \inf_{\Omega}\left\{\frac{\mathrm{Vol}(\partial\Omega)}{\mathrm{Vol}(M)} \;\middle|\; \Omega \subset M,\ \mathrm{Vol}(\Omega) = \beta\mathrm{Vol}(M)\right\}, \text{ for } \beta$$

in $[0,1]$ (it should be clear that $\mathrm{Vol}(\Omega)$ is an n-dimensional volume, $n = \dim M$, and that $\mathrm{Vol}(\partial\Omega)$ is an $(n-1)$-dimensional volume), and $\Omega$ smooth domains in $M$.

An isoperimetric estimator on $(M,g)$ is a function $H$ from $[0,1]$ to $\mathbb{R}_+$ such that $h(\beta) \geq H(\beta)$ for all $\beta$ in $[0,1]$.

10. Example

Let us consider $(S^2,\mathrm{can})$. The volume of a geodesic ball of radius $r$ in $(S^2,\mathrm{can})$ is $2\pi(1-\cos r)$, and the volume of the corresponding geodesic sphere of radius $r$ is $2\pi \sin r$. It follows from the isoperimetric inequality on $(\$^n_k,\mathrm{can})$ (see end of §A) that $h(S^2,\mathrm{can};\beta) = \sqrt{\beta(1-\beta)}$.

11. Proposition

The function $h(M,g;\beta)$ has the following properties ($n = \dim M$)

(i) $h(\beta) \geq 0$;

(ii) $h(\beta) = h(1-\beta)$;

(iii) $h(\beta) \sim C^*(n)\mathrm{Vol}(M)^{-1/n}\,\beta^{(n-1)/n}$, when $\beta$ is close to 0 (for $C^*(n)$ see formula (8));

(iv) $h(M,g;.)$ is continuous on $[0,1]$ and has right and left derivatives at each $\beta$ in $]0,1[$ and is differentiable in $]0,1[$ except on a denumerable set.

Proof. (i) and (ii) are clear;

(iii) is called the asymptotic isoperimetric inequality; it says that for domains with small volume, the isoperimetric inequality looks very much like the classical isoperimetric inequality in $\mathbb{R}^n$: see [B-M] Appendix C;

(iv) is much more delicate: see [B-B-G2] ■

12. Philosophy

Isoperimetric function vs. Isoperimetric inequality.

If we want to use isoperimetric methods on a given Riemannian manifold $(M,g)$, the best thing we can hope for is to know the isoperimetric function $h(M,g;\beta)$ itself. In general this is not the case and we have to replace $h(\beta)$ by some minorizing function $H(\beta)$.

Now if we want to use isoperimetric methods on a class of Riemannian manifolds, we have to choose an isoperimetric inequality which is valid for any manifold in the given class.

An example of such a class of Riemannian manifolds is

$$\mathfrak{M}_{n,k,D} = \{(M,g) \mid \dim M = n,\ \mathrm{Ricci}(M,g) \geq (n-1)kg,\ \mathrm{Diam}(M,g) \leq D\}.$$

We then consider isoperimetric inequalities of the form $h(M,g;\beta) \geq H(n,k,D;\beta)$, for any $(M,g)$ in $\mathfrak{M}_{n,k,D}$. We will give examples of such situations in Chapters V and VI.

In the sequel we will study isoperimetric methods on Riemannian manifolds with a given isoperimetric estimator $H(\beta)$ and with the above philosophy in mind.

## 13. Symmetrization

Let $(M,g)$ be a Riemannian manifold equipped with an isoperimetric estimator $H(\beta)$. Because a generic Riemannian manifold does not have symmetries, we cannot compare a domain in $M$ with a geodesic ball in $M$. Keeping in mind the symmetrization procedure which we used in the proof of the Faber-Krahn inequality, we will instead construct a model space with nice balls having isoperimetric properties related to $H(\beta)$.

For this purpose we consider $S^{n-1} \times ]0,L[$ with Riemannian metric $g^* = a^2(s)d\theta^2 + ds^2$, where $\theta$ is in $S^{n-1}$, $s$ in $]0,L[$, and $d\theta^2$ is the canonical Riemannian metric on $(S^{n-1},\mathrm{can})$. We also assume that $a(0) = a(L) = 0$. We call $(M^*,g^*)$ this Riemannian manifold (it is not necessarily complete; we also use $M^*$ for $S^{n-1} \times ]0,L[ \cup \{N,S\}$, where the north and south poles $N$ and $S$ are the points corresponding to $S^{n-1} \times \{0\}$ and $S^{n-1} \times \{L\}$; for $M^*$ to be smooth one needs that $a'(0) = 1$ and $a'(L) = -1$; $(M^*,g^*)$ can be viewed as a manifold with revolution symmetry).

We denote the volume of $(M^*,g^*)$ by $V^*$. We call $A(s)$ the relative volume of the ball $B(N,s)$ with center $N$ and radius $s$ (i.e. $B(N,s) = \{N\} \cup S^{n-1} \times ]0,s[$). We then have

$$(14) \qquad A(s) = V^{*-1}\,\mathrm{Vol}(S^{n-1}) \int_0^s a^{n-1}(t)dt.$$

(see Fig. 5).

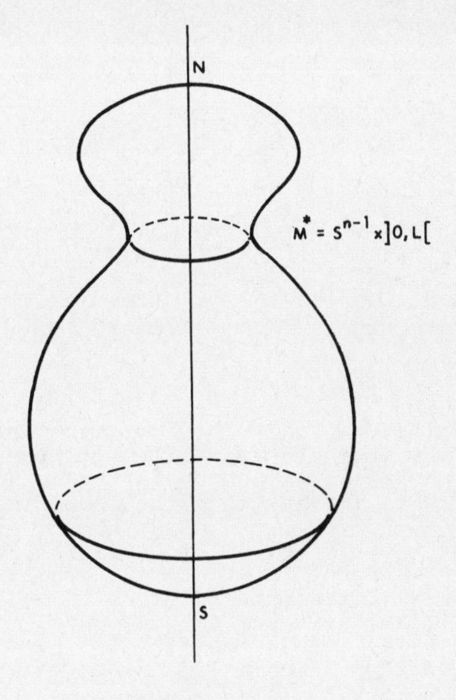


Fig.5

15. In order to have a nice symmetrization procedure on the manifold $(M,g;H(\beta))$, we want to construct a model $(M^*,g^*)$ such that the balls $B(N,s)$ have $H$ as isoperimetric function i.e. such that

$V^{*-1}\,\mathrm{Vol}(\partial B(N,s)) = H(A(s))$, (recall that $A(s)$ is the relative volume of $B(N,s)$).

This can also be written as

(16) $A'(s) = H(A(s))$, $s \in ]0,L[$, because $\mathrm{Vol}(\partial B(N,s)) =$ $= \mathrm{Vol}(S^{n-1})a^{n-1}(s)$.

Given a Riemannian manifold $(M,g)$, with isoperimetric estimator $H$, we can construct $(M^*,g^*)$ as follows. We determine $A(s)$ by the differential equation (16) and the initial condition $A(0) = 0$ (we could also use another condition: see nº 22 below).

This determines $A(s)$ by the equality

$$(17) \qquad s = \int_0^{A(s)} \frac{du}{H(u)} ,$$

from which we deduce the value of $L$

$$(18) \qquad L = \int_0^1 \frac{du}{H(u)} .$$

19. Remark

It is clear that the isoperimetric function $h$ satisfies $h(0) = h(1) = 0$. This implies that the isoperimetric estimator $H$ must satisfy $H(0) = H(1) = 0$. It follows that Equations (17) and (18) only make sense if the integrals converge. Notice that in view of Proposition 11 (iii), the integral $\int_0^1 \frac{du}{h(u)}$ converges. For this reason, we will usually assume that $H(\beta) \sim C\beta^\alpha$ when $\beta$ is close to $0$ with $1 > \alpha \geq \frac{n-1}{n}$ and a similar assumption near $\beta = 1$ (the second inequality comes from Proposition 11 (iii) and the assumption that $h(\beta) \geq H(\beta)$ for all $\beta$). We shall also give an example with $\alpha = 1$ (see nº 22 below).

20. So far, given $(M,g;H(\beta))$, we have determined $A(s)$ and $L$. In order to determine $(M^*,g^*)$, we still have a degree of freedom, namely the choice of $V^*$. This will turn out to be convenient later (nº 21 and Chap. V) but the choice of $V^*$ is in fact irrelevant for the following reasons. Let $\tilde{a}(s) = V^{*-1/(n-1)}a(s)$. This function is determined by $A(s)$ because

$$A'(s) = V^{*-1}\mathrm{Vol}(S^{n-1})a^{n-1}(s) = \mathrm{Vol}(S^{n-1})\,\tilde{a}^{n-1}(s).$$

Now let $f$ be a function on $M^*$ which depends only on the variable $s$. The Rayleigh quotient $R(f)$ of $f$ on $(M^*,g^*)$ is given by

$$R(f) = \int_{S^{n-1}}\int_0^L \dot{f}^2(s)a^{n-1}(s)dvds \Big/ \int_{S^{n-1}}\int_0^L f^2(s)a^{n-1}(s)dvds,$$

where $dv$ is the Riemannian measure on $(S^{n-1},\mathrm{can})$. It follows that

$$R(f) = \int_0^L \dot{f}^2(s)\,\tilde{a}^{n-1}(s)ds \Big/ \int_0^L f^2(s)\tilde{a}^{n-1}(s)ds$$

does not depend on $V^*$: this is a Rayleigh quotient in one dimension, with measure $\tilde{a}^{n-1}(s)ds$. In the sequel we will compare the Rayleigh quotient of a function on $(M,g)$ to the Rayleigh quotient of a radial function on $(M^*,g^*)$ (i.e. depending only on the s-variable) so that we will be able to ignore $V^*$.

On the other hand, the radial part of the Laplacian on $(M^*,g^*)$ is given by (see nº III.5(v))

$$-\frac{\partial^2}{\partial s^2} - (n-1)\frac{a'(s)}{a(s)}\frac{\partial}{\partial s}$$

(because $\theta(s,u) = a^{n-1}(s)$ in the local chart $\exp_N$).
This operator does not depend on a choice of $V^*$.

As a matter of fact, it turns out that our manifold with revolution symmetry $(M^*,g^*)$ is just a convenient way of visualizing a <u>one-dimensional model</u>.

21. <u>Example</u>

Let $(M,g)$ be a Riemannian surface with isoperimetric

estimator $H(\beta) = \sqrt{\beta(1-\beta)}$. Formula (17) gives

$$s = \int_0^{A(s)} [u(1-u)]^{-1/2} du, \quad \text{i.e.} \quad A(s) = \sin^2 \frac{s}{2} \text{ and } L=\pi.$$

It then follows that $a(s) = \frac{V^*}{4\pi} \sin s$, and that

$$(M^*, g^*) = (S^1 \times ]0,\pi[,\ g^* = a^2(s)d\theta^2 + ds^2).$$

A pleasant choice of $V^*$ is $V^* = 4\pi$, in which case $(M^*,g^*)$ is just $(S^2,\text{can})$, whose isoperimetric function is $h(S^2,\text{can};\beta) = \sqrt{\beta(1-\beta)}$ (see Example 10). Another choice of $V^*$ gives a "cigar" with two conic points and constant curvature 1.

22. Application: Cheeger's isoperimetric inequality.

In 1970, J. Cheeger introduced the following isoperimetric constant, known as Cheeger's isoperimetric constant. For a closed Riemannian manifold $(M,g)$ we define

$$h_C = h_C(M,g) = \inf\{\text{Vol}(\partial\Omega)/\text{Vol}(\Omega) \mid \Omega \subset M,\ 2\text{Vol}(\Omega) \leq \text{Vol}(M)\},$$

$\Omega$ smooth domains in $M$. It follows that

$$h(\beta) = h(M,g;\beta) \geq h_C \min(\beta, 1-\beta),$$

so that we can choose $H(\beta) = h_C \min(\beta,1-\beta)$. Notice that $H(\beta)=H(1-\beta)$, $\int_0^1 \frac{du}{H(u)}$ diverges at 0 and 1 (see Remark 19). Taking into account the symmetry of $H(\beta)$, we construct the model space $(M^*,g^*)$ as $S^{n-1} \times ]-\infty\infty[$ with $g^* = a^2(s)d\theta^2 + ds^2$, and we solve the differential equation (16) with the initial condition $A(0) = 1/2$, which gives

$$s = \int_{1/2}^{A(s)} \frac{du}{H(u)}. \tag{23}$$

Using the symmetry of H, we find that for <u>positive</u> s, $A(s) + A(-s) = 1$ and hence $a(s) = a(-s)$. Finally (23) gives

For $s \geq 0$,

$$(24) \qquad \begin{cases} A(s) = 1 - \frac{1}{2}\exp(-h_C s) \\ \\ a(s) = (V^* \mathrm{Vol}(S^{n-1})^{-1} h_C/2)^{1/(n-1)} \exp(-h_C s/(n-1)) \end{cases}$$

so that $(M^*, g^*)$ is made of two "cusps" glued together in a symmetric manner (see Fig. 6 in dimension 2).

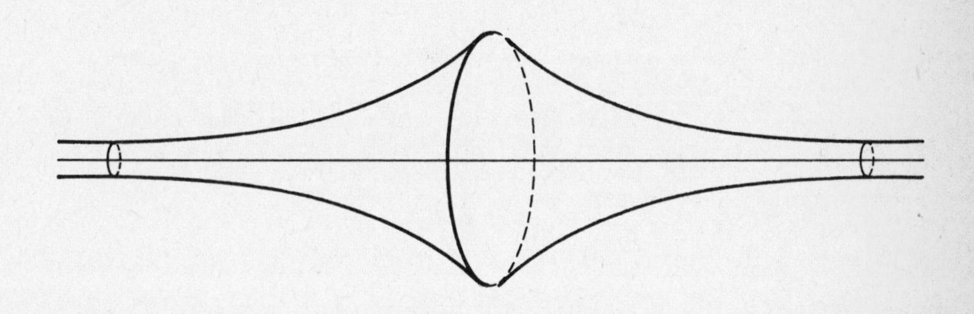

Fig. 6

The manifold $(M^*, g^*)$ is complete, noncompact and has finite volume. Using the symmetry of $M^*$, we can view N as "the point at $\infty$ : $S^{n-1} \times \{\infty\}$".

25. Cheeger's isoperimetric constant was introduced in order to give a lower bound for $\lambda_2(M,g;\ \text{closed})$ (see nº III.19). The first

eigenvalue $\lambda_1(M,g;\text{ Closed})$ is always 0, the corresponding eigenspace corresponding to constant functions. Cheeger proved that

$$(26) \qquad \lambda_2(M,g;\text{ Closed}) \geq h_C^2/4.$$

An eigenfunction $f$ associated with $\lambda_2(M)$ being orthogonal to the constants must change sign in $M$. We can therefore find a connected component $\Omega$ of $M\setminus f^{-1}(0)$, such that $2\text{Vol}(\Omega) \leq \text{Vol}(M)$. We can also assume that $f$ is positive in $\Omega$. It turns out that $f|\Omega$ is a first eigenfunction associated with the first eigenvalue $\lambda_1(\Omega,g;\text{ Dirichlet})$ of the Dirichlet eigenvalue problem in $\Omega$, and that $\lambda_1(\Omega,g;D) = \lambda_2(M,g;C)$ (think of Green's formula, Theorem III.10 or see [B-M] Appendix D).

We will now estimate $\lambda_1(\Omega,g;D)$ from below in terms of Cheeger's isoperimetric constant $h_C$. For this purpose we use the model $(M^*,g^*)$ of Fig. 6 above, with $V^* = \text{Vol}(M)$ and we mimic the proof of the Faber-Krahn inequality (§A).

Let $\Omega_t = \{x \text{ in } \Omega \mid f(x) \geq t\}$ and let $\Omega_t^* = S^{n-1}\times[r(t),\infty[\subset M^*$ be such that $\text{Vol}(\Omega_t) = \text{Vol}(\Omega_t^*)$. Define a function $\phi\colon [r(0),\infty[\to \mathbb{R}_+$ by $\phi(r(t)) = t$. The function $\phi$ increases from 0 to $\sup f = m$ in $[r(0),\infty[$. For $(\theta,s)$ in $\Omega^* = \Omega_o^*$, let $f^*(\theta,s) = \phi(s)$. The co-area formula gives (see § A)

$$\int_\Omega |df|^2 dv_g = \int_0^m \Big( \int_{G(t)} |df|\, da_t\Big) dt,$$

where $G(t) = (f|\Omega)^{-1}(t)$. By Cauchy-Schwarz inequality, we can write

$$\int_{G(t)} |df|\, da_t \geq \Big(\int_{G(t)} da_t\Big)^2 \Big/ \int_{G(t)} |df|^{-1} da_t .$$

Since $2\text{ Vol }\Omega_t \leq \text{Vol}(M)$, we have

$$\text{Vol}(\partial\Omega_t) \geq h_C\text{Vol}(\Omega_t) = h_C\text{Vol}(\Omega_t^*) = \text{Vol}(\partial\Omega_t^*),$$

by definition of $(M^*, g^*)$. This can be written as

$$\mathrm{Vol}(G(t)) \geq \mathrm{Vol}(G^*(t))$$

where $G^*(t) = f^{*-1}(t)$. Since $f^*$ only depends on the s-variable, we can write

$$\int_{G(t)} |df|\, da_t \geq \int_{G^*(t)} |df^*|\, da_t^*$$

(we again used the fact that $\mathrm{Vol}\,\Omega_t = \mathrm{Vol}\,\Omega_t^*$: see § A).

As in § A we conclude that

$$(27) \qquad \int_\Omega |df|^2 dv_g \Big/ \int_\Omega f^2 dv_g \geq \int_{\Omega^*} |df^*|^2 dv_{g^*} \Big/ \int_{\Omega^*} f^{*2}\, dv_{g^*}.$$

The right-hand side of Inequality (27) can be written as

$$\int_{r(0)}^\infty \dot{\phi}^2(s) \exp(-h_C s) ds \Big/ \int_{r(0)}^\infty \phi^2(s) \exp(-h_C s) ds = R_1(\phi),$$

because $g^* = a^2(s) d\theta^2 + ds^2$ (note that $r(0) \geq 0$ and that $|df^*|$ is the norm of $df^*$ on $M^*$ for the dual metric $g^{*-1}$). It follows that

$$\lambda_2(M, g; \text{Closed}) = \int_\Omega |df|^2 dv_g \Big/ \int_\Omega f^2 dv_g \geq \inf\{R_1(\phi)\} = \Lambda,$$

where the infimum in the right-hand side is taken over all functions $\phi$ such that

(i) $\phi$ and $\dot{\phi}$ are in $L^2(\mathbb{R}_+, \exp(-h_C s) ds)$,

(ii) $\phi(0) = 0$

($\dot{\phi}$ the derivative of $\phi$ is the sense of distributions).

It is easy to see that

$$\Lambda = \inf\{R_1(\phi) \mid \phi \text{ in } C_o^\infty(\mathbb{R}_+^\bullet)\},$$

and that

$$\Lambda = \inf_{n>0} \inf\{R_1(\phi) \mid \phi \text{ in } C_o^\infty(]0,n[)\}.$$

Using the first variational characterization of the eigenvalues nº III.26), it follows that

$$\Lambda_n = \inf\{R_1(\phi) \mid \phi \text{ in } C_o^\infty(]0,n[)\}$$

is the first eigenvalue of the Dirichlet eigenvalue problem

$$\begin{cases} \phi''(s) - h_C\phi'(s) + \lambda\phi(s) = 0 \\ \phi(0) = \phi(n) = 0 \end{cases}$$

which can be solved explicitly showing that $\Lambda_n \geq h_C^2/4$. Finally we can conclude that $\Lambda \geq h_C^2/4$, which proves Cheeger's estimate (26).

28. Remarks.

(i) Cheeger's original proof is shorter than the above one. Although it uses the same technical details as that of the Faber-Krahn inequality, it is quite different. We found it interesting to show that Cheeger's inequality can be reduced to an inequality à la Faber-Krahn, with an appropriate model space $(M^*,g^*)$;

(ii) One can also consider the surface with boundary $S^1\times[0,\infty[$, with the above metric $g^*$. This manifold is not compact but is complete with finite volume. One can still consider the Laplacian $\Delta$ as an unbounded operator on $L^2(M^*,g^*)$ with Dirichlet boundary condition. The number $h_C^2/4$ then appears as the lower bound of the spectrum of the Friedrichs extension of $\Delta$ (compare with [CL]

Chap. IV.3) (here $\Delta$ has a continuous spectrum see [R-S]);

(iii) Cheeger's estimate (26) would be void of sense if we did not know that $h_C > 0$. In fact one can prove the following estimate (see [GA1]).

29. Theorem. *Let* $(M,g)$ *be an* n-*dimensional compact Riemannian manifold without boundary. Define* $r_{min}$ *by*

$r_{min} = \inf\{Ricci(M,g)(u,u) \mid u \text{ in } UM\}$, *where* UM *is the unit tangent bundle to* M.

*Assume that* $r_{min}.Diam(M,g)^2 \geq \varepsilon(n-1)k^2$, $\varepsilon \in \{-1,0,1\}$, $k \in \mathbb{R}_+^*$.

Then

$$Diam(M,g).h_C(M,g) \geq K(k), \quad \textit{where}$$

$$K(k) = \begin{cases} k\left[\int_0^{k/2} (\cos t)^{n-1}dt\right]^{-1} & \textit{if} \quad \varepsilon = 1 \\ 2 & \textit{if} \quad \varepsilon = 0 \\ k\left[\int_0^{k/2} (\operatorname{ch} t)^{n-1}dt\right]^{-1} & \textit{if} \quad \varepsilon = -1 \end{cases}.$$

30. Remarks.

(i) Notice that both $r_{min}.Diam(M,g)^2$ and $Diam(M,g).h_C(M,g)$ are Riemannian invariants of weight 0 (see nº II.6);

(ii) Theorem 29 shows that $h_C(M,g)$ is uniformly bounded from below on the class $\mathfrak{M}_{n,k,D}$ given in nº 12;

(iii) In Chapter VI we will give an estimate on $\lambda_2(M,g;C)$ which is sharper than Cheeger's and we will generalize this estimate to $\lambda_i(M,g;C)$, $i \geq 2$.

31. Comments.

(i) Let $(M,g)$ be a Riemannian manifold (always assumed to be compact without boundary) equipped with an isoperimetric estimator $H(\beta)$. If $\int_0^1 \frac{du}{H(u)}$ converges, the manifold $M^*$ is compact (possibly with two conic points) and we can easily mimic the proof of the Faber-Krahn inequality to show the following assertion.

Let $\Omega$ be a domain in $M$ and let $\Omega^*$ be the ball $B(N,r)$ in $M^*$ such that $Vol(\Omega)/Vol(M) = Vol(\Omega^*)/Vol(M^*)$. Then

$$(32) \qquad \lambda_1(\Omega,g;D) \geq \lambda_1(\Omega^*,g^*;D).$$

Notice that in order to find $\lambda_1(\Omega^*,g^*;D)$, one only has to solve a one-dimensional eigenvalue problem (indeed the first eigenfunction is radial: compare with nº 20).

As was already pointed out in Remark 28 (iii), the estimate (32) is void if we do not know $H(\beta)$ i.e. if we cannot give lower bounds for $h(\beta)$ in terms of geometric data. So again the main difficulty is to find a good isoperimetric inequality. This fact will turn out to be even more important in Chapter V: see nº V § C;

(ii) One can also investigate isoperimetric inequalities on a manifold with boundary. In the case of a domain $\Omega$ in a manifold without boundary $M$, we have used the isoperimetric inequality in $M$ to obtain results on the Dirichlet eigenvalue problem in $\Omega$. One can also consider isoperimetric constants adapted to the Dirichlet boundary condition. For example one can define Cheeger's isoperimetric constant

$$(33) \qquad h_C(\Omega,g;\text{Dirichlet}) = \inf\{Vol(\partial\omega)/Vol(\omega) \mid \omega \subset \mathring{\Omega}\}$$

for the Dirichlet boundary conditions on $\partial\Omega$. If we want to deal with the Neumann problem, we have to allow subdomains $\omega$ such that $\partial\omega \cap \partial\Omega \neq \phi$; see [BR] p. 29. It turns out that the isoperimetric constants adapted to the Neumann boundary conditions are much more difficult do deal with than the other ones. In fact, estimates on $\lambda_1(\Omega, g;\text{Neumann})$ involve the geometry of $(\Omega, \partial\Omega)$ is a very strong way. We shall not deal with these problems here: see [ME1] for more details.

For further reading on Cheeger's constant $h_C$ we recommend [BR].

<u>Further references for Chapter IV</u>: [BE], [CL] Chap. IV, [PE], [P-S], [ON].

# CHAPTER V

# ISOPERIMETRIC METHODS AND THE HEAT EQUATION

ALL RIEMANNIAN MANIFOLDS ARE ASSUMED TO BE

COMPACT, CONNECTED, WITHOUT BOUNDARY

In Chapter I, we used the wave equation and separation of variables to motivate eigenvalue problems; we could have used the heat equation as well.

In the present chapter, we give direct results concerning the heat kernel of a Riemannian manifold. They will be useful in Chapter VI.

## A. THE HEAT EQUATION.

Let $(M,g)$ be a compact Riemannian manifold without boundary. To determine the heat flow $u(t,x)$ on $(M,g)$ is to find a function $u(t,x)$, solution of the heat equation:

$$(1)\qquad \begin{cases} \dfrac{\partial u}{\partial t}(t,x) + \Delta_x u(t,x) = f(t,x), \text{ for } (t,x) \text{ in } \mathbb{R}_+^{\bullet} \times M, \\ \\ u(0,x) = f_o(x), \text{ for } x \text{ in } M, \end{cases}$$

where $f_o$ and $f$ are given functions (e.g. $C^\infty$ functions).

An easy way to solve this problem is to introduce the notion of fundamental solution of the heat equation (or heat kernel) on $(M,g)$. The heat kernel is a function $k$ on $\mathbb{R}_+^{\bullet} \times M \times M$ which satisfies the following properties

$$(2)\quad \begin{cases} \text{(i)} & k(t,x,y) \text{ is continuous on } \mathbb{R}_+^{\bullet} \times M \times M,\ C^1 \text{ in the } t\text{-} \\ & \text{variable and } C^2 \text{ in the } x\text{-variable;} \\ \text{(ii)} & (\frac{\partial}{\partial t} + \Delta_x)k(t,x,y) = 0 \text{ for all } (t,x,y) \text{ in } \mathbb{R}_+^{\bullet} \times M \times M; \\ \text{(iii)} & \lim_{t\to 0} k(t,x,y) = \delta_x(y), \text{ the Dirac measure at } x. \end{cases}$$

Property (2.iii) means that for any $h$ in $C^\infty(M)$, we have

$$\lim_{t\to 0_+} \int_M k(t,x,y)h(y)dv_g(y) = h(x),$$

and is usually written as $k(0,x,y) = \delta_x(y)$.

At least at the formal level, the solution $u(t,x)$ of (1) is given by the following formula (known as Duhamel's formula)

$$u(t,x) = \int_M k(t,x,y)f_o(y)dv_g(y) + \int_0^t (\int_M k(t-s,x,y)f(s,y)dv_g(y))ds.$$

For the following theorem we refer to [CL] Chap. VI, [B-G-M] Chap. III.E., or [GY].

3. *Theorem. Let* $(M,g)$ *be an n-dimensional compact Riemannian manifold without boundary, with eigenvalues (counted with multiplicities)* $\{\lambda_i\}_{i\geq 1}$ *and associated orthonormal real eigenfunctions* $\{\phi_i\}_{i\geq 1}$. *There exists a unique heat kernel* $k(t,x,y)$ *on* $(M,g)$. *This is a* $C^\infty$ *function on* $\mathbb{R}_+^{\bullet} \times M \times M$ *which satisfies* $k(t,x,y) = k(t,y,x)$ *for all* $(t,x,y)$ *in* $\mathbb{R}_+^{\bullet} \times M \times M$. *Furthermore*, $k(t,x,y)$ *can be expressed as*

$$k(t,x,y) = \sum_{j=1}^{\infty} \exp(-\lambda_j t)\phi_j(x)\phi_j(y),$$

*where the series in the right-hand side converges in the* $C^k$*-topology* on any subset of the form $[a,\infty[ \times M \times M$, $a > 0$, for any $k$.

For example, this theorem justifies Duhamel's formula. Theorem 3 also justifies the following equalities

$$(4) \quad \begin{cases} \text{(i)} & k(t,x,x) = \sum_{j=1}^{\infty} \exp(-\lambda_j t)\phi_j^2(x) \\ \text{(ii)} & Z(t) = \sum_{j=1}^{\infty} \exp(-\lambda_j t) = \int_M k(t,x,x)dv_g(x). \end{cases}$$

The function $Z(t)$ (the trace of the heat kernel on $(M,g)$) is called the partition function of $(M,g)$. We shall also use the notation $Z(M,g;t)$ to stress the dependence of $Z(t)$ on $(M,g)$.

5. Exercise. Prove that giving the sequence of eigenvalues (with multiplicities) $\{\lambda_i\}_{i\geq 1}$ of $(M,g)$ is equivalent to giving the partition function $Z(t)$ of $(M,g)$.

6. Examples and Exercises.

(i) Although $(\mathbb{R}^n, \text{can})$ is not a compact Riemannian manifold, it has a heat kernel: $(4\pi t)^{-n/2} \exp(-\|x-y\|^2/4t)$; however, we cannot take its trace as we did for $k(t,x,y)$ in (4);

(ii) Let $\Gamma$ be a lattice in $\mathbb{R}^n$ and let $\Gamma^*$ be the dual lattice: $\Gamma^* = \{\gamma^* \in \mathbb{R}^n \mid \text{for all } x \in \Gamma, \langle x|\gamma^*\rangle \in \mathbb{Z}\}$. The eigenvalues of the torus $T_\Gamma = (\mathbb{R}^n/\Gamma, \text{can}/\Gamma)$ are the numbers $4\pi^2\|\gamma^*\|^2$, with associated orthonormal complex eigenfunctions $\exp(2i\pi\langle x|\gamma^*\rangle)\text{Vol}(T_\Gamma)^{-1/2}$, $\gamma^* \in \Gamma^*$. The heat kernel of $T_\Gamma$ is given by

$$k(t,x,y) = (4\pi t)^{-n/2} \sum_{\gamma\in\Gamma} \exp(-\|x-y-\gamma\|^2/4t).$$

In particular, Formula (4ii) can be written as

$$(4\pi t)^{-n/2}\mathrm{Vol}(T_\Gamma) \sum_{\gamma\in\Gamma} \exp(-\|\gamma\|^2/4t) = \sum_{\gamma\in\Gamma^*} \exp(-4\pi^2\|\gamma^*\|^2 t)$$

(this formula is known as Poisson summation formula).

It follows from the Poisson summation formula that $Z(T_\Gamma,\mathrm{can};t) \sim (4\pi t)^{-n/2}\mathrm{Vol}(T_\Gamma)$ when $t$ goes to $0_+$ ($n=\dim T_\Gamma$). In fact the following property holds.

7. *Property*. *For any n-dimensional Riemannian manifold* $(M,g)$

$$Z(M,g;t) \sim (4\pi t)^{-n/2}\mathrm{Vol}(M,g) \quad \textit{when} \quad t \to 0_+.$$

For more details see Chap. VII or [CL] Chap. VI, [B-G-M] Chap. III.E.

The purpose of this chapter is to give an isoperimetric inequality for the heat kernel.

## B. ISOPERIMETRIC INEQUALITY FOR THE HEAT KERNEL, I.

8. Let $(M,g)$ be an n-dimensional Riemannian manifold (compact, without boundary) and assume that one is given an isoperimetric estimator $H$ on $(M,g)$. As in Chapter IV, we construct a model space $(M^*,g^*)$ which is associated with $H$ (and with a choice of $V^*$). We assume that $H$ satisfies

$$H(\beta) \sim C\beta^\alpha, \; 1 > \alpha \geq \frac{n-1}{n},$$

for $\beta$ close to $0$ and a similar property when $\beta$ is close to $1$ (see nº IV.19). This implies that $(M^*,g^*)$ is a "nice Riemannian manifold" possibly with two conic points (the north and south poles);

recall that although we see a manifold with revolution symmetry, the mathematics only see a one-dimensional model: see nº IV.20.

Our main theorem is the following (see [B-G] § 2, [B-B-G1] §III).

9. *Theorem. Under the above assumptions* (nº 8), *let* $k(t,x,y)$ *denote the heat kernel of the Riemannian manifold* $(M,g)$ *and let* $k_*(t,N,N)$ *denote the heat kernel of the Riemannian manifold* $(M^*,g^*)$ *evaluated at* $(t,N,N)$, *where* $N$ *is the north pole of* $(M^*,g^*)$. *The following inequalities hold*

$$Z(M,g;t) \leq \mathrm{Vol}(M,g) \sup_x k(t,x,x) \leq \mathrm{Vol}(M^*,g^*) k_*(t,N,N).$$

10. The main ideas in the proof of Theorem 9 are as follows (compare with [BE] Chap. IV §3, [M-T])

(a) We consider Problem (1) with $f \equiv 0$ and $f_o > 0$ on $(M,g)$, and we compare the solution $u(t,x)$ to the solution of a *symmetrized problem* on $(M^*,g^*)$;

(b) We apply a symmetrization procedure similar to the one described in Chapter IV, but we take the relative volume as new parameter;

(c) We then let $f_o$ tend to a Dirac measure on $(M,g)$, so that we obtain a comparison theorem for $k(t,x,y)$.

11. Since we are mainly interested in geometry in these notes, we only give a rough sketch of the proof of Theorem 9. For full analytic details see [BE] IV.3, [B-G], [M-T].

We divide the proof into several steps.

12. <u>Step 1</u>. Let $f$ be a $C^\infty$ positive function on $M$. We define

$$D(r) = \{x \in M \mid f(x) > r\} \quad \text{and}$$

$$a(r) = \mathrm{Vol}(D(r))/\mathrm{Vol}(M).$$

We now define a function $\bar{f}$ by

$$\bar{f}(s) = \inf\{r \mid a(r) < s\}.$$

The function $a(r)$ is non-increasing, and varies from $a(0) = 1$ to $0$, when $r$ increases from $0$ to $\sup f$. If $a$ were strictly decreasing and continuous, $\bar{f}$ would be the inverse function of $a$ (see [TI] Chap. I). Since $f$ is $C^\infty$, it follows from Sard's theorem that $a$ is $C^\infty$ on an open set whose complement has measure zero (use the co-area formula nº IV.5). We also have $\bar{f}(a(r)) = r$ for all regular values $r$. See Fig. 7

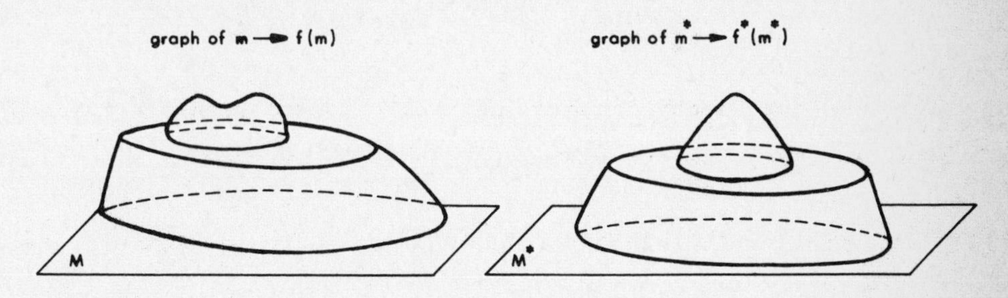


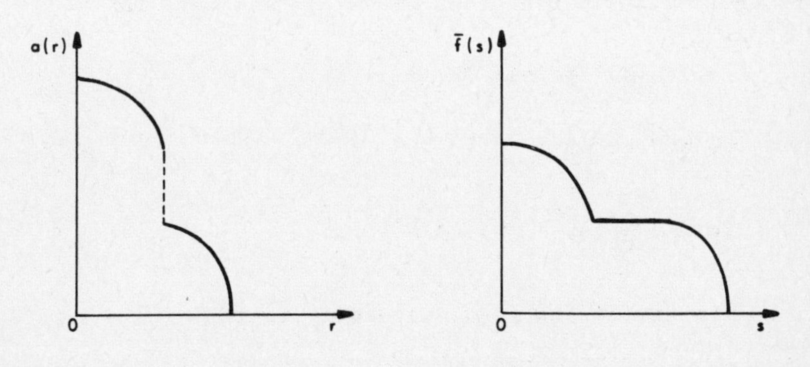


Fig. 7

Let $E(s) = D(\bar{f}(s))$, $G(s) = \partial E(s)$ and

$$F(s) = \int_{E(s)} f(x)dv_g(x).$$

It follows from the co-area formula (nº IV.5) that

$$\mathrm{Vol}(M)a'(r) = -\int_{\partial D(r)} |df|^{-1}da_r,$$

when $r$ is a regular value of $f$. We deduce that

$$(13) \qquad F(s) = \mathrm{Vol}(M)\int_0^s \bar{f}(u)du.$$

Since the sets $D(r)$ are level sets of $f$, we deduce from Green's formula (Theorem III.10) that

$$\int_{E(s)} \Delta f(x)dv_g(x) = \int_{G(s)} \frac{\partial f}{\partial \nu}\, da_s = \int_{G(s)} |df|da_s.$$

Applying Cauchy-Schwarz inequality to the righ-hand side of the above equality, we can write

$$(14) \qquad \int_{E(s)} \Delta f dv_g \geq (\mathrm{Vol}G(s))^2/\int_{G(s)} |df|^{-1}da_s = -(\mathrm{Vol}G(s))^2/\mathrm{Vol}(M)a'(\bar{f}(s)).$$

It follows from the definitions of a, $\bar{f}$ and H, that

$$\mathrm{Vol}(G(s)) \geq \mathrm{Vol}(M)H(s).$$

From (13) we deduce that

$$\frac{d^2F}{ds^2}(s) = \mathrm{Vol}(M)\,\frac{df}{ds}(s) = \mathrm{Vol}(M)/a'(\bar{f}(s)).$$

From these relations and (14), we deduce that

$$(15) \qquad \int_{E(s)} \Delta f\, dv_g \geq -H^2(s)\,\frac{d^2F}{ds^2}(s).$$

Let $f^\wedge$ be a radial $C^\infty$ function on $(M^*,g^*)$. With the obvious notations and taking into account the fact that $f^\wedge$ is radial, and the very definition of $M^*$ we have

$$(16) \qquad \int_{E^\wedge(s)} \Delta^* f^\wedge \, dv_{g^*} = -H^2(s)\, \frac{d^2F^\wedge}{ds^2}(s)$$

(where $\Delta^*$ is the Laplacian on $(M^*,g^*)$).

17. Step 2. Let $f_o$ be a positive $C^\infty$ function on M. Let $u(t,x)$ be the solution of Problem (1) (with $f \equiv 0$). It follows from the maximum principle for the heat equation ([P-W]) that $u(t,x)$ is positive.

Fixing $t$, we can apply the first step to $u(t,.)$. We define

$$a_t(r) = \mathrm{Vol}\{x \in M \mid u(t,x) > r\}/\mathrm{Vol}(M),$$

$$\bar{u}_t(s) = \inf\{r \mid a_t(r) < s\},$$

$$E(t,s) = \{x \in M \mid u(t,x) > \bar{u}_t(s)\},$$

$$F(t,s) = \int_{E(t,s)} u(t,x)dv_g(x).$$

We deduce from the first step that

$$\int_{E(t,s)} \Delta u(t,x)dv_g(x) \geq -H^2(s)\, \frac{\partial^2 F}{\partial s^2}(t,s).$$

18. Lemma. ([BE] Lemma 4.23 p. 212)

$$\int_{E(t,s)} \frac{\partial u}{\partial t}(t,x)dv_g(x) = \frac{\partial F}{\partial t}(t,s).$$

Finally, we conclude from Lemma 18 and the preceding inequality (recall that $u(t,x)$ solves the heat equation) that

(19) $$\frac{\partial F}{\partial t}(t,s) - H^2(s)\frac{\partial^2 F}{\partial s^2}(t,s) \leq 0.$$

In particular, by letting $f_o$ tend to the Dirac measure $\delta_y$ at $y$ in $M$ we conclude that we can also take $u(t,x)=k(t,x,y)$ ($y$ fixed).

20. Let $f_o^\wedge$ be a $C^\infty$ radial decreasing function in $(M^*,g^*)$. We consider the solution $u^\wedge(t,x)$ of

$$\left(\frac{\partial}{\partial t} + \Delta^*\right)u^\wedge(t,x) = 0,$$

$$u^\wedge(0,x) = f_o^\wedge(x),$$

on $(M^*,g^*)$. We conclude that $u^\wedge(t,x)$ is also a $C^\infty$-radial decreasing function on $(M^*,g^*)$ ([BE] Prop. 4.8 p.214).

If we apply to $u^\wedge$ what we did before for $u$ and if we take the first step into account, we conclude (using obvious notations) that

(21) $$\frac{\partial F^\wedge}{\partial t}(t,s) - H^2(s)\frac{\partial^2 F^\wedge}{\partial s^2}(t,s) = 0.$$

Now we choose a sequence of radial decreasing function $f_{o,n}^\wedge$ converging to $\delta_N$, the Dirac measure at $N$ in $(M^*,g^*)$. It follows that (21) also holds for $k_*(t,N,.) = u^\wedge(t,.)$ where $k_*$ is the heat kernel on $(M^*,g^*)$.

22. <u>Step 3</u>. Using $u(t,x) = k(t,x,y)$, $y$ fixed in $M$, and $u^\wedge(t,x) = k_*(t,N,x)$ we define $h(t,s)$ by

$$h(t,s) = F(t,s) - F^\wedge(t,s).$$

This function satisfies the following properties.

23. <u>Properties</u>.

(i) $\frac{\partial h}{\partial t}(t,s) - H^2(s)\frac{\partial^2 h}{\partial s^2}(t,s) \le 0$ <u>for</u> $(t,s)$ <u>in</u> $\mathbb{R}_+^\bullet \times [0,1]$;

(ii) $h(t,0) = 0$ <u>for all</u> $t > 0$ (<u>we integrate functions on a set with volume equal to</u> 0);

(iii) $\lim_{t\to 0_+} h(t,s) = 0$ <u>for all</u> $s \in [0,1]$ (<u>because</u> $k$ <u>and</u> $k_*$ <u>are heat kernels</u>);

(iv) $h(t,1) = 0$ <u>for all</u> $t > 0$ ($\frac{\partial h}{\partial t}(t,1) = 0$ <u>for all</u> $t$, <u>because</u> $\int_M \Delta f \, dv_g = 0$ <u>for all</u> $C^\infty$ $f$ <u>and</u> $h(0,1) = 0$).

From these properties and the maximum principle (see [P-W]) applied to $h$, we conclude that $h(t,s) \le 0$ or

$$(24)\quad \begin{cases} \text{For all } (t,s) \text{ in } \mathbb{R}_+^\bullet \times [0,1],\ F(t,s) \le F^\wedge(t,s), \text{ or equivalently,} \\ \text{For all } (t,s)\ \mathrm{Vol}(M)\displaystyle\int_0^s \bar{u}_t(r)dr \le \mathrm{Vol}(M^*)\int_0^s \bar{u}_t^\wedge(r)dr. \end{cases}$$

It follows from the convexity of $t \to t^2$ and from the second mean value theorem ([BE] p. 173-174), that

$$(25)\quad \text{For all } t > 0,\ \mathrm{Vol}(M)^2\int_0^1 \bar{u}_t^2(r)dr \le \mathrm{Vol}(M^*)^2\int_0^1 \bar{u}_t^{\wedge 2}(r)dr.$$

Now, $\mathrm{Vol}(M)\int_0^1 \bar{u}_t^2(r)dr = \int_M k^2(t,x,y)dv_g(y) = k(2t,x,x)$,

where the second equality follows from the <u>semi-group property</u> of the heat kernel $k$ on $M$ (e.g. use Theorem 3) and similarly $\mathrm{Vol}(M^*)\int_0^1 \bar{u}_t^{\wedge 2}(r)dr = k_*(2t,N,N)$. Finally we have proved that

$$\mathrm{Vol}(M)k(2t,x,x) \le \mathrm{Vol}(M^*)k_*(2t,N,N),$$

From which Theorem 9 easily follows ■

C. ISOPERIMETRIC INEQUALITY FOR THE HEAT KERNEL, II.

As we already mentionned in relation with Cheeger's estimate in Chapter IV nº 28 (iii), Theorem 9 is only interesting if we have a "good" isoperimetric estimator $H(\beta)$ on $(M,g)$. As we also pointed out in nº IV.12, it is not always possible to use the isoperimetric function $h(\beta)$ itself, although its properties (Proposition IV.11) allow us to construct a model $(M_h^*, g_h^*)$ with a "good" heat kernel (see nº 29 below).

The following theorem (see [B-B-G1]) gives a nice isoperimetric inequality for heat kernel comparisons (for another Theorem see [B-G] p. XV.17).

26. Theorem. *Let $(M,g)$ be an n-dimensional compact Riemannian manifold without boundary. We define*

$$r_{min}(M) = \inf\{\mathrm{Ricci}(M,g)(u,u) \mid u \text{ unit tangent vector to } M\},$$

$$d(M) = \mathrm{Diam}(M,g).$$

*If $(M,g)$ satisfies $r_{min}(M)d(M)^2 \ge \varepsilon(n-1)\alpha^2$ for $\varepsilon \in \{-1,0,1\}$ and $\alpha \in \mathbb{R}_+^*$, there exists a positive number $a(n,\varepsilon,\alpha)$ such that for all $\beta$ in $[0,1]$,*

$$h(M,g;\beta) \ge h(S^n(R), \mathrm{can};\beta)$$

*where $S^n(R)$ is the sphere of radius $R = d(M)/a(n,\varepsilon,\alpha)$, in $\mathbb{R}^{n+1}$ with induced metric.*

The number $a(n,\varepsilon,\alpha)$ is defined by

$$a(n,\varepsilon,\alpha) = \begin{cases} \alpha\omega_n^{1/n}\left(2\int_0^{\alpha/2}\cos^{n-1}(t)dt\right)^{-1/n} & \text{if } \varepsilon = 1; \\ (1 + n\omega_n)^{1/n} - 1 & \text{if } \varepsilon = 0; \\ \alpha c(\alpha) & \text{if } \varepsilon = -1, \end{cases}$$

where $\omega_n = \mathrm{Vol}(S^n)/\mathrm{Vol}(S^{n-1})$, and where $c(\alpha)$ is the unique positive root of the equation

$$x\int_0^{\alpha}(\mathrm{ch}\,t + x\,\mathrm{sh}\,t)^{n-1}dt = \omega_n.$$

The proof of this theorem is rather difficult, we refer to [B-B-G1].

27. Remarks.

(a) When $\varepsilon = 1$, Myers' Theorem (nº II.29) implies that $\alpha \le \pi$;

(b) Myers' theorem also shows that the estimate when $\varepsilon = 1$ improves Gromov's isoperimetric inequality on manifolds with positive Ricci curvature (see [GV1]); Theorem 26 generalizes Gromov's theorem to all manifolds, whatever the sign of their Ricci curvature;

(c) In the case $\varepsilon = -1$, we can also replace $a(n,-1,\alpha)$ by the following lower bound for $\alpha c(\alpha)$ (see [B-B-G1]) $\alpha c(\alpha) \ge \alpha \min\{C(\alpha), C(\alpha)^{1/n}\}$, where

$$C(\alpha) = (n-1)\omega_n/[\exp((n-1)\alpha) - 1];$$

(d) If we now take $H(M,g;\beta) = h(S^n(R),\mathrm{can};\beta)$ as isoperimetric estimator on $(M,g)$, we notice that $H(\beta) = H(1-\beta)$ and that $H(\beta) \sim C\beta^{(n-1)/n}$ when $\beta$ tends to $0$ (for some constant $C$). The model space $(M^*,g^*)$ associated with $H(\beta)$ is then $(S^n(R),\mathrm{can})$.

Taking into account the behaviour of $k_M(t,x,y)$ under scaling (e.g. use Theorem 3) and the fact that $(S^n, can)$ is 2-point homogeneous, which implies that $k_{S^n}(t,x,x)$ is independent of $x$, we deduce from Theorem 9 and Theorem 26 the following (see [B-B-Gl] §III).

28. Theorem. Under the assumptions of Theorem 26, we have

$$Z(M,g;t) \le Vol(M,g) \sup_M k_M(t,x,x) \le Z(S^n, can; t/R^2)$$

(where $R = d(M)/a(n,\varepsilon,\alpha)$).

29. Remarks. (of a philosophical flavor) In Theorem 9, we gave a comparison theorem using the heat kernel $k_*$ on $(M^*, g^*)$. As the proof of Theorem 9 shows, we only used the function $K_*(t,x) =$ $= k_*(t,N,x)$ on $(M^*, g^*)$. The function $K_*(t,.)$ is a radial function with respect to $N$, so that we only use the radial part of the heat kernel, associated with the radial part of the Laplacian (see nº IV.20)

$$\Delta_r^* = -\{\frac{\partial^2}{\partial r^2} + (n-1)\frac{a'(r)}{a(r)}\frac{\partial}{\partial r}\},$$

where $(M^*, g^*) = (S^{n-1} \times ]0,L[, a^2(r)d\theta^2 + dr^2)$.

Assume that $H(\beta) \sim C\beta^{\alpha}$, $1 > \alpha \ge (n-1)/n$, when $\beta$ tends to $0$ (for some constant $C$). It is then easy to check (by making an appropriate choice for $V^*$) that

$$a(r) \sim r \quad \text{if} \quad \alpha = (n-1)/n, \quad \text{and}$$

$$a(r) \sim r^{\gamma}, \quad \gamma > 1 \quad \text{if} \quad 1 > \alpha > (n-1)/n,$$

when $r$ tends to $0$.

Now recall that the radial part of the Laplacian in $(\mathbb{R}^n, can)$ is

$$\Delta_r^n = -\{\frac{\partial^2}{\partial r^2} + \frac{n-1}{r}\frac{\partial}{\partial r}\}$$ (this non-standard notation will only be used in these few lines).

We conclude that for $r$ close to $0$, $\Delta_r^*$ looks like $\Delta_r^n$ when $\alpha = \frac{n-1}{n}$ and looks like $\Delta_r^p$, $p > n$ $(p \in \mathbb{R})$, when $1 > \alpha > \frac{n-1}{n}$. This means that our comparison function $k_*(t,N,N)$ will look like an "n-dimensional" heat kernel if we choose $\alpha = \frac{n-1}{n}$, and like a "p-dimensional" heat kernel $(p > n)$ when $1 > \alpha > \frac{n-1}{n}$. In particular, its behaviour when $t$ tends to $0_+$ will be in $t^{-n/2}$ (resp. $t^{-p/2}$) when $\alpha = \frac{n-1}{n}$ (resp. $1 > \alpha > \frac{n-1}{n}$).

Because of Proposition IV.11, we see that it is much better to take an isoperimetric estimator $H$ such that $\alpha = \frac{n-1}{n}$; this is the case if we use Theorem 26.

Another interpretation can be made in terms of the behaviour of the function $A(s)$ (see nº IV.14) when $s$ goes to $0$. If $A(s) \sim C\, s^m$ when $s$ goes to $0$, the volume of a small geodesic ball in $(M^*,g^*)$ is of the order of $s^m$ and hence $M^*$ has "isoperimetric" dimension $m$ (*) (recall that we are near a conic point). It is clear that it is better to compare $(M,g)$ to an m-dimensional manifold, with $m = \dim M$. This means again that we have to take $\alpha = \frac{n-1}{n}$.

The case $\alpha = 1$ is even worst because $(M^*,g^*)$ is no longer compact (see nº IV.22).

Let us also mention here that the isoperimetric function $h(\beta)$ has exactly the required properties which allow us to define the heat kernel for $\Delta_r^*$ on the associated model space (see Proposition IV.11 or [B-B-G2]).

---

* This is a non-standard notion which we use here for convenience.

## D. APPLICATIONS

In this paragraph we give some direct applications of Theorem 28; for further results we refer to [B-G] §3 or [B-B-G1] § III.

30. Let $(M,g)$ be a compact Riemannian manifold without boundary such that (<u>notations as in</u> nº 26-29)

$$n = \dim M;$$

$$r_{min}(M)d(M)^2 \geq \varepsilon(n-1)\alpha^2.$$

Let $\{\lambda_i\}_{i\geq 1}$ be the sequence of eigenvalues (counted with multiplicities) of $(M,g)$ [Notice that we count the eigenvalues from $i = 1,\ldots$, some authors begin with $i = 0$ e.g. [B-G-M]].

31. <u>Theorem</u>. <u>Let</u> $(M,g)$ <u>be as above. Then</u>

(i) $\lambda_2(M,g) \geq a^2(n,\varepsilon,\alpha)\, n\, d(M)^{-2}$;

(ii) <u>there exists a number</u> $C(n,\varepsilon,\alpha)$ <u>such that for</u> $i \geq 2$,

$$\lambda_i(M,g) \geq C(n,\varepsilon,\alpha)\, i^{2/n}\, d(M)^{-2}.$$

<u>Proof</u>. Using Theorem 28, we can write

$$(*)\ Z(M,g;t) \leq Z(S^n,\mathrm{can};a^2(n,\varepsilon,\alpha)t\, d(M)^{-2}).$$

For any Riemannian manifold (compact, connected, without boundary), we have

$$Z(M,g;t) = 1 + \sum_{j=2}^{\infty} \exp(-\lambda_j(M,g)t)$$

because $\lambda_1(M,g) = 0$ has multiplicity 1.

Assertion (i) follows from (*) by substracting 1 to both

sides, by taking Log and by letting $t$ tend to infinity.

It follows from Property 7 that there exists constants $C(n)$ and $D(n)$ such that

$$(**) \text{ for all } t > 0,\ Z(S^n,\mathrm{can};t) \le C(n)t^{-n/2} + D(n).$$

Let $N(\lambda) = \mathrm{Card}\{j \mid \lambda_j(M,g) \le \lambda\}$. We can then write

$$k \le N(\lambda_k) \le e \sum_{\lambda_j \le \lambda_k} \exp(-\lambda_j/\lambda_k) \le eZ(M,g;1/\lambda_k)$$

and, using Theorem 28 and $(**)$,

$$k \le e\, C(n)(d(M)/a(n,\varepsilon,\alpha))^n \lambda_k^{n/2} + e\, D(n)$$

Assertion (ii) follows ■

32. Remarks.

(i) A theorem of Lichnerowicz states that if $r_{min}(M) \ge (n-1)$ then $\lambda_2(M,g) \ge n = \lambda_2(S^n,\mathrm{can})$. Recall that Myers' theorem implies that $d(M) \le \pi$. The expression of $a(n,\varepsilon,\alpha)$ in Theorem 26 together with Theorem 31(i) give

$$r_{min}(M) \ge n-1 \Rightarrow \lambda_2(M,g) \ge$$

$$\ge n \left\{ \int_0^{\pi/2} \cos^{n-1}(t)dt \Big/ \int_0^{d(M)/2} \cos^{n-1}(t)dt \right\}^{2/n} > n$$

when $d(M) < \pi$. In fact, one can show that $\lambda_2(M,g) = n$ implies that $(M,g)$ is isometric to $(S^n,\mathrm{can})$; this is Obata's theorem: see [B-G-M] Chap. III.D and compare with [CG] and [CL] Chap. III.4. In fact Assertion (i) in Theorem 31 can also be proved by an argument à la Faber-Krahn (see nº IV.31); the Lichnerowicz-Obata theorem also follows from this method: see [B-M];

(ii) For $r_{min}(M) \geq 0$, Theorem 31(i) and Theorem 26 give

$\lambda_2(M,g) \geq 8/d(M)^2$ when $\dim M = 2$; on the other hand, $\lambda_2(S^1(R) \times S^1) \sim \pi^2/d^2(S^1(r) \times S^1)$ when $r$ goes to infinity;

(iii) Weyl's estimate gives

$\lambda_k(M,g) \sim C(n)k^{2/n}\,\mathrm{Vol}(M,g)^{-2/n}$ when $k$ goes to infinity. It turns out that one cannot substitute $\mathrm{Vol}(M,g)^{-2/n}$ to $d(M)^{-2}$ in Theorem 31(i), as the following example shows. Consider the Riemannian manifolds

$$M_a = S^{n-1}(a^{1/(n-1)}) \times S^1(1/a)$$

with the product metric; they satisfy $\mathrm{Vol}(M_a) = \mathrm{Vol}(S^{n-1}) \times \mathrm{Vol}(S^1) = \mathrm{Vol}(M_1)$, and $\mathrm{Diam}(M_a)$ goes to infinity when a goes to zero. The number $N_a(\lambda)$ of eigenvalues of $M_a$ less than $\lambda$ satisfies $N_a(\lambda) \geq 2\,\mathrm{Card}\{p \in \mathbb{N}^\bullet \mid a^2p^2 \leq \lambda\}$. This shows that for fixed $\lambda, N_a(\lambda)$ goes to infinity when a goes to zero. In particular, this implies that $\lambda_k(M_a)\mathrm{Vol}(M_a)^{2/n}$ goes to zero with a. However it follows from Theorem 31(ii) that $\lambda_k(M_a)d(M_a)^2$ is bounded from below when a goes to zero.

For a counter-example involving the lower bound on Ricci see [GA2], I.1.2.

At least qualitatively, the estimate in Theorem 31(ii) is best possible (it was obtained by Gromov in [GV1] with a worst constant).

33. Remark. Since $k_M(t,x,x) = \sum_{j=1}^{\infty} \exp(-\lambda_j t)\phi_j^2(x)$, Theorem 28 also gives bounds on the $L^\infty$-norm of the eigenfunctions $\phi_j$ of $\Delta$ on $(M,g)$.

The next chapter is devoted to inverse geometric results. We will use Theorem 28 in a crucial way.

Further references for Chapter V: For other comparison theorems on the heat kernel see [CL] or the references in [B-B]. For the heat kernel itself, see [DK], [GY].

# CHAPTER VI

## GEOMETRIC APPLICATIONS OF ISOPERIMETRIC METHODS

ALL RIEMANNIAN MANIFOLDS ARE COMPACT,
CONNECTED WITHOUT BOUNDARY

In this chapter, we will give some partial answers to the following geometric inverse problem.

1. Problem. To what extent do local estimates on the curvature of a Riemannian manifold $(M,g)$ inforce global restrictions on the manifold?

### A. INTRODUCTION.

In order to explain the meaning of Problem 1, let us give an appropriate formulation of the Gauss-Bonnet theorem ([HF] Part II, Chap. III).

2. Theorem. *Let $(M,g)$ be any compact Riemannian surface, whose curvature $K$ is bounded from below by the real number $k$. Then*

(i) $$\chi(M) = \frac{1}{2\pi}\int_M K(M)dv_g \geq \frac{k}{2\pi}\mathrm{Vol}(M,g);$$

(ii) $$b_1(M) \leq 2 - \frac{k}{2\pi}\mathrm{Vol}(M,g).$$

*Here* $\chi(M)$ *denotes the Euler characteristic of* $M$ *and* $b_1(M) = 2 - \chi(M)$ *the first Betti number of* $M$ *(these are topological invariants which do not depend on the choice of a Riemannian metric* $g$ *on* $M$); $\mathrm{Vol}(M,g)$ *is the* 2-*dimensional volume of* $(M,g)$.

3. *Corollary. The number of differentiable surfaces which admit a Riemannian metric whose curvature is bounded from below by* $k$, *and whose volume is bounded from above by* $V$ $(k$ *in* $\mathbb{R}$, $V$ *in* $\mathbb{R}_+)$ *is finite*.

4. *Comments*.

(i) Let us first point out that the product $K(M)\mathrm{Vol}(M)$ is a Riemannian invariant with weight $0$ in dimension 2;

(ii) By scaling the metric, it is always possible to bound any curvature of an n-dimensional manifold by 1 in absolute value, so that we cannot expect any general theorem answering Problem 1, without scaling. In order to scale the metric, we can use a Riemannian invariant, e.g. the volume or the diameter. In the Gauss-Bonnet theorem, the metric is scaled by giving an upper bound on the volume. In *general*, we will have to *use the diameter*; the following example shows that fixing the volume is a very weak condition. Take any manifold $(N,g)$. By an appropriate choice of $R$, the Riemannian manifold $(N \times S^1(R), g \times \mathrm{can}_R) = (M,g_R)$ has volume one. However the topology of $M$ may be very complicated;

(iii) In dimension bigger than 2, we have several notions of curvature. We will always try to use the weakest possible notion. In general we will try to use the *Ricci curvature* (the scalar curvature is very often too weak an invariant). This is the case if we want bounds on the eigenvalues of the Laplacian $\Delta^g$ of $(M,g)$: see nº V.31. In other situations, we will have to make assumptions on the sectional curvature (see nº 24 (ii)).

Finally, we reduce Problem 1 to the following.

5. Problem. Give global bounds on $M$ (e.g. on topological invariants) in terms of (Diam(M,g) and Ricci(M,g)) or (Diam(M,g) and Sect(M,g)).

6. Examples.

(i) Myers' theorem (see nº II.29) says that if Ricci(M,g) > 0, then $\pi_1(M)$, the fundamental group of $M$, is finite. This is a partial answer to Problem 5. Notice that no scaling is required here. In fact, a consequence of Myers' theorem is that $\mathrm{Diam}(M,g) \leq \pi/k$, if $\mathrm{Ricci}(M,g) \geq (n-1)k^2 > 0$. However, taking Diam(M,g) into account gives sharper results (see nº V.32 and [B-B-Gl] Corollary 17), so that in some sense scaling is also necessary here;

(ii) In the 1940's, S. Bochner proved the following results

$$\mathrm{Ricci}(M,g) > 0 \Rightarrow b_1(M) = 0 \quad \text{(1st Betti number)},$$

$$\mathrm{Ricci}(M,g) \geq 0 \Rightarrow b_1(M) \leq \dim(M).$$

These results were obtained by an analytic method which we now describe (notice that Myers' theorem is proved by geometric methods).

B. THE ANALYTIC APPROACH, I.

7. Let $(M,g)$ be an n-dimensional Riemannian manifold (compact, connected, without boundary). We denote by $\Lambda^p T^*M$, $0 \leq p \leq n$, the $p^{th}$ exterior product of $T^*M$ and by $E^p(M)$ the $C^\infty$ sections of

$\wedge^p T^* M$, i.e. the exterior forms of degree $p$ on $M$. The exterior differential $d$ is a first order differential operator from $E^p(M)$ to $E^{p+1}(M)$. This operator only depends on the differentiable structure.

We now define an operator $\delta: E^{p+1}(M) \to E^p(M)$ by

$$\langle\langle \alpha | \delta\beta \rangle\rangle = \langle\langle d\alpha | \beta \rangle\rangle, \tag{8}$$

for all $\alpha$ in $E^p(M)$ and $\beta$ in $E^{p+1}(M)$: the metric $g$ on $TM$ induces a metric on each $\wedge^p T_x^* M$ which we denote by $\langle . | . \rangle_x$; we define $\delta$ by

$$\int_M \langle \alpha | \delta\beta \rangle_x dv_g(x) = \int_M \langle d\alpha | \beta \rangle_x \, dv_g(x)$$

(note that in the first integral we have the scalar product of two p-forms, and in the second integral the scalar product of two $(p+1)$-forms). We say that $\delta$ is the <u>formal adjoint</u> of $d$ (note that $\delta$ depends on the Riemannian metric $g$).

9. <u>Properties</u>.

(i) <u>If</u> $f \in C^\infty(M) = E^\circ(M)$, <u>then</u> $\delta f = 0$;

(ii) <u>If</u> $\alpha \in E^1(M)$, <u>then</u> $\delta\alpha = -\mathrm{Div}_g(\alpha^\#)$;

(iii) <u>Let</u> $\{e_1, \ldots, e_n\}$ <u>be a local orthonormal frame in</u> $M$. <u>For</u> $\alpha$ <u>in</u> $E^p(M)$ <u>we note</u>

$\alpha(i_1, \ldots, i_p)$ <u>for</u> $\alpha(e_{i_1}, \ldots, e_{i_p})$ <u>where</u> $i_1, \ldots, i_p \in \{1, \ldots, n\}$.

<u>Then</u>

$$d\alpha(i_1, \ldots, i_{p+1}) = \sum_{k=1}^{p+1} (-1)^{k+1} (D_{i_k}\alpha)(i_1, \ldots, \hat{i}_k, \ldots, i_{p+1}),$$

$$\delta\alpha(i_2,\ldots,i_p) = -\sum_{k=1}^{n} (D_k\alpha)(k,i_2,\ldots,i_p)$$

(<u>recall that we note</u> $D_k$ <u>for</u> $D_{e_k}$, <u>where</u> D <u>is the Levi-Civita connection of</u> $(M,g)$);

(iv) <u>If</u> M <u>is oriented and if</u> $*$ <u>denotes the Hodge operator on</u> M, $*: E^p(M) \to E^{n-p}(M)$, <u>then</u>

$\delta: E^p(M) \to E^{p-1}(M)$ <u>satisfies</u>

$$\delta = (-1)^{n(p+1)} * d * .$$

<u>Proof</u>.

(i) is dual to the fact that $d\alpha = 0$ for all $\alpha$ in $E^n(M)$;

(ii) follows from the definitions of $\delta$ and $\mathrm{Div}_g$ (see also [B-G-M] Chap. II.G);

(iii) the formulae for d and $\delta$ follow from the definitions of d, $\delta$ and of the Levi-Civita connection D of $(M,g)$;

For (iv) see [WA] Chap. 6. [The formula $\langle\alpha|\beta\rangle_x dv_g(x) = \beta\wedge*\alpha$, defines $*$ uniquely] ■

10. <u>Definition</u>. One defines the <u>Laplace-Beltrami operator</u> (or Laplacian) on p-forms by $\Delta = \delta d + d\delta$ (this operator is also called Hodge-de Rham Laplacian).

For further details on $\delta,\Delta$ see [WA] Chap. 6 and [CL] Appendix, or [LZ] and [RM].

The classical Hodge- de Rham theory ([WA] Chap. 5 and 6) states that

(11) $\qquad b_p(M) = \dim \mathrm{Harm}^p(M);$

the $p^{th}$ Betti number of the manifold $M$ is equal to the dimension of the space of harmonic p-forms $(\mathrm{Harm}^p(M) = \{\alpha \in E^p(M) \mid \Delta\alpha=0\})$.

Note that $b_p(M)$ is a topological invariant, while $\mathrm{Harm}^p(M)$ depends on the Riemannian metric.

In order to prove Bochner's results, and to introduce the analytic method, we need the following.

12. <u>Lemma</u>. <u>Let</u> $(M,g)$ <u>be a Riemannian manifold and let</u> $\alpha$ <u>be a</u> 1-<u>form on</u> $M$. <u>The following formulae hold</u>

(i) $\Delta\alpha = D^*D\alpha + \mathrm{Ricci}(\alpha^\#,.)$;

(ii) $\langle\Delta\alpha|\alpha\rangle = \frac{1}{2}\,\Delta(\langle\alpha|\alpha\rangle) + |D\alpha|^2 + \mathrm{Ricci}(\alpha^\#,\alpha^\#)$.

In Formula (i), the Laplacian is the Laplacian acting on 1-forms, $D$ the Riemannian connection on 1-forms, and $D^*$ its adjoint; for a tensor field $\beta$, $D^*$ is given by $D^*\beta = -\mathrm{Trace}\, D\beta$ (contraction of the first two indices or $D^*\beta(.) = -\sum_k D_k\beta(k,.)$, in a local orthonormal frame (notations as in nº 9). Equivalently, we have

$$D^*D\alpha = -\sum_{i=1}^{n}\{D_i(D_i\alpha) - D_{D_i i}\alpha\},$$

in a local orthonormal frame (notations as in nº 9).

In formula (ii), the left hand side is the point-wise scalar product of two 1-forms; $\Delta(\langle\alpha|\alpha\rangle)$ is the Laplacian of the function $\langle\alpha|\alpha\rangle$ and $|D\alpha|$ is the norm of the 2-tensor $D\alpha$.

13. <u>Definitions</u>. The operator $\bar{\Delta} = D^*D$ is called the <u>rough Laplacian</u> (here on 1-forms). Formulae (i) and (ii) are called <u>Weitzenböck formulae</u>.

Proof of Lemma 12. We use the notations of nº 9; we denote by $\{e_i\}$ a local orthonormal frame near $x$; we can always assume that we have $(D_{e_i} e_i)(x) = 0$, at the point $x$ (see III.40).

Claim 1. For $\beta$ a section of $\otimes^p T^*M$, we have

$$D^*\beta = -\text{Trace } D\beta = -\sum_k D_k\beta(k,.).$$

Let $\gamma$ be a section of $\otimes^{p-1}T^*M$. We consider the 1-form $\omega = \sum_I \beta(.,I)\gamma(I)$, where $I$ is a multi-index of length $(p-1)$. Now $\delta\omega$ is a function on $M$ which is given at $x$ by (see nº 9 (iii))

$\delta\omega = -\sum_{k=1}^{n} e_k . (\sum_I \beta(k,I)\gamma(I))$. An easy computation gives

$$\delta\omega(x) = -\langle D^{\wedge}\beta|\gamma\rangle_x + \langle\beta|D\gamma\rangle_x$$

where $D^{\wedge}\beta = -\text{Trace } D\beta$. Since this is valid for all $x$ in $M$, and since $\int_M \delta\omega \, dv_g = 0$ (Divergence Theorem III.9), we have $\langle\langle D^{\wedge}\beta|\gamma\rangle\rangle = \langle\langle\beta|D\gamma\rangle\rangle$, which shows that $D^{\wedge} = D^*$ (see nº III.9 and [B-G-M] Chap. II.GII).

Claim 2. For any 1-form $\alpha$,

$$D^*D\alpha = -\sum_{i=1}^{n} \{D_i(D_i\alpha) - D_{D_i i}\alpha\}.$$

Using the first claim we have for all $k$, $D^*D\alpha(e_k) = -\sum_j D_j\beta(j,k)$, where $\beta = D\alpha$, i.e.

$$D^*D\alpha(e_k) = -\sum_j e_j . \beta(j,k) + \sum_j[\beta(D_j j,k) + \beta(j,D_j k)].$$

Now $\beta(j,k) = D\alpha(j;k) = (D_j\alpha)(k)$, so that

$$D^*D\alpha(e_k) = -\sum_j D_j((D_j\alpha))(k) - \sum_j (D_j\alpha)(D_jk)$$

$$+ \sum_j (D_{D_jj}\alpha)(k) + \sum_j (D_j\alpha)(D_jk)$$

$$= -\sum_j \{D_j(D_j\alpha) - D_{D_jj}\alpha\}(k).$$

Note that at the point $x$ we can write

$$D^*D\alpha_x = - \sum_{i=1}^{n} D_i(D_i\alpha)_x \quad \text{since} \quad D_i i(x) = 0.$$

Proof of formula (i)

We use the formula

$$d\alpha(i,j) = (D_i\alpha)(j) - (D_j\alpha)(i), \quad \text{see 9(iii)}$$

(this formula easily follows from the definitions of $d$ and $D$:

$$d\alpha(X,Y) = X.\alpha(Y) - Y.\alpha(X) - \alpha([X,Y])).$$

This gives, at the point $x$ $(D_i j(x) = 0)$,

$$\delta d\alpha(i)_x = - \sum_{k=1}^{n} e_k.(d\alpha(e_k,e_i))_x$$

$$= - \sum_{k=1}^{n} D_k(D_k\alpha)(i)_x + \sum_{k=1}^{n} e_k.((D_i\alpha)(k))_x.$$

We can also write

$$d\delta\alpha(i)_x = e_i.(\delta\alpha) = - \sum_{k=1}^{n} e_i.(D_k\alpha(k)).$$

Finally we can write

$$\Delta\alpha(i)_x = \bar{\Delta}\alpha(i)_x + \sum_{k=1}^{n} \{e_k.((D_i\alpha)(k))_x - e_i.((D_k\alpha)(k))_x\}.$$

Since $(D_j\alpha)^\# = D_j(\alpha^\#)$, the second term in the right-hand side can be written as

$$\langle (D_kD_i - D_iD_k)\alpha^\#,\ e_k\rangle_x = \langle R(e_k,e_i)\alpha^\# + D_{[e_k,e_i]}\alpha^\#,\ e_k\rangle_x,$$

in view of nº II.43. Now we have $[e_k,e_i](x) = 0$ (see nº II.35) so that

$$\Delta\alpha(i)_x = \bar{\Delta}\alpha(i)_x + \sum_{k=1}^{n} \langle R(e_k,e_i)\alpha^\#,e_k\rangle_x$$

$$= \bar{\Delta}\alpha(i)_x + \sum_{k=1}^{n} R(e_k,e_i;e_k,\alpha^\#)_x \quad \text{(see nº II.45)}$$

$$= \bar{\Delta}\alpha(i)_x + \mathrm{Ricci}_x(\alpha^\#,e_i) \quad \text{(see nº II.45).}$$

Proof of formula (ii)

In order to prove (ii), it suffices to prove

$$\langle \bar{\Delta}\alpha|\alpha\rangle = \frac{1}{2}\,\Delta(\langle\alpha|\alpha\rangle) + |D\alpha|^2.$$

We can write

$$e_k.\langle\alpha|\alpha\rangle = 2\langle D_k\alpha|\alpha\rangle, \quad \text{and}$$

$$\Delta(\langle\alpha|\alpha\rangle)_x = -\sum_k e_k.(e_k(\langle\alpha|\alpha\rangle))_x$$

$$= -2\sum_k\langle D_k\alpha|D_k\alpha\rangle_x - 2\sum_k\langle D_k(D_k\alpha)|\alpha\rangle_x$$

($D_k i = 0$ at $x$). Finally we have

$$\Delta(\langle\alpha|\alpha\rangle)_x = -2|D\alpha|_x^2 + 2\langle\Delta\alpha|\alpha\rangle_x \quad \blacksquare$$

14. Definition. A 1-form $\alpha$ is parallel if $D\alpha = 0$.

## 15. Exercises.

(i) Show that the point-wise norm $|\alpha|_x$ of a parallel 1-form $\alpha$ is constant;

(ii) Show that the vector-space of parallel 1-forms has dimension less than or equal to $n = \dim M$ [Hint: Take a curve $c(t)$ in $M$ and let $\{e_i\}$ be a parallel orthonormal frame along $c$ (see nº II.40). Show that if $\alpha_{c(t)} = \Sigma\, \alpha_i(t) e_i^\flat(t)$, then the $\alpha_i(t)$ are constant functions].

## 16. Proof of Bochner's results.

Integrating formula 12(ii) on $M$, we obtain

$$\langle\langle \Delta\alpha | \alpha \rangle\rangle = \int_M |D\alpha|^2 \, dv_g + \int_M \mathrm{Ricci}(\alpha^\#, \alpha^\#) dv_g,$$

because $\int_M \Delta f \, dv_g = 0$ for any $C^\infty$ function $f$.

In view of the Hodge- de Rham theory (nº (11)), we take $\alpha$ to be a harmonic 1-form. Finally we obtain

(17) Any harmonic 1-form $\alpha$ satisfies

$$\int_M |D\alpha|^2 dv_g + \int_M \mathrm{Ricci}(\alpha^\#, \alpha^\#) dv_g = 0.$$

For $x$ in $M$, we define

$$(18)\quad \begin{cases} \rho(x) = \inf\{\mathrm{Ricci}(u,u) \mid u \text{ unit vector in } T_x M\}, \text{ and} \\ r_{min} = \inf\{\rho(x) \mid x \text{ in } M\}. \end{cases}$$

We have the inequalities

$$(19)\qquad \int_M \mathrm{Ricci}(\alpha^\#,\alpha^\#)dv_g \geq \int_M \rho(x)\langle\alpha|\alpha\rangle_x dv_g \geq r_{min}\ \langle\langle\alpha|\alpha\rangle\rangle.$$

20. Proposition. With the above notations, we have the following results.

(i) If $\rho(x) \geq 0$, and if there exists an $x_o$ such that $\rho(x_o) > 0$, then $b_1(M) = 0$;

(ii) If $r_{min} = 0$, then $b_1(M) \leq n = \dim M$.

Proof. Assumption (i) implies that $\int_M |D\alpha|^2\ dv_g < 0$ for any non-zero harmonic 1-form: this is impossible, hence $b_1(M) = 0$.

Assumption (ii) implies that $\int |D\alpha|^2 dv_g \leq 0$ and hence that $D\alpha = 0$. Any harmonic 1-form is parallel and hence $b_1(M) \leq n$, by Exercise 15(ii) ■

Notice that Assertion (i) in the Proposition is sharper than Bochner's result as stated in nº 6 (ii).

21. Generalizations.

The above situation can be generalized as follows. We consider a fiber bundle $E$ over the Riemannian manifold $(M,g)$. We assume that $E$ is equipped with a Riemannian metric (we say that $E$ is a Riemannian fibre bundle) i.e. with a scalar product $\langle.|.\rangle_x$ in the fibers $E_x$ of $E$, depending $C^\infty$ on $x$. We also assume that $E$ is equipped with a connection $\bar{D}$, which is compatible with the scalar product i.e.

$$\bar{D}: \mathfrak{A}(M) \times C^\infty(M) \to C^\infty(E)$$

is an $\mathbb{R}$-linear map which satisfies the following properties

(i) $X.\langle u|v\rangle = \langle \bar{D}_X u|v\rangle + \langle u|\bar{D}_X v\rangle$,

for all $X$ in $\mathfrak{U}(M)$ and $u,v$ in $C^\infty(E)$;

(ii) $\bar{D}_{fX}u = f\,\bar{D}_X u$,

for all $f$ in $C^\infty(M)$, $X$ in $\mathfrak{U}(M)$ and $u$ in $C^\infty(E)$;

(iii) $\bar{D}_X(fu) = (X.f)u + f\,\bar{D}_X u$,

for all $f$ in $C^\infty(M)$, $X$ in $\mathfrak{U}(M)$ and $u$ in $C^\infty(E)$.

Finally, we assume that there is a natural Laplacian $\tilde{\Delta}$ acting on $C^\infty(E)$ (i.e. a $2^{nd}$ order linear partial differential operator, with properties similar to those of the Laplace-Beltrami operator) and that $\tilde{\Delta}$ satisfies the following <u>Weitzenböck formula</u>

(22) $\tilde{\Delta}s = \bar{\Delta}s + \mathfrak{R}s$,

where $\bar{\Delta}s = -\sum_{i=1}^{n}\{\bar{D}_{e_i}(\bar{D}_{e_i}s) - \bar{D}_{D_{e_i}e_i}s\}$ is the <u>rough Laplacian</u>, $\{e_i\}$ local orthonormal frame in $M$, and where $\mathfrak{R}$ is a <u>symmetric endomorphism</u> of the bundle $E$, $\mathfrak{R}_x\colon E_x \to E_x$ is an endomorphism of $E_x$ which satisfies $\langle \mathfrak{R}_x u|v\rangle_x = \langle u|\mathfrak{R}_x v\rangle_x$ for all $u,v$ in $E_x$.

As above we define

$$(23)\quad \begin{cases} \mathfrak{R}(x) = \inf\{\langle \mathfrak{R}_x u|u\rangle_x \mid u \text{ in } E_x,\ \langle u|u\rangle_x = 1\}, \\ \mathfrak{R}_{min} = \inf\{\mathfrak{R}(x) \mid x \text{ in } M\}. \end{cases}$$

24. <u>Examples</u>.

(i) $E = T^*M$, $\bar{D}$ is the Levi-Civita connection, $\tilde{\Delta}$ is the Laplacian on 1-forms, $\mathfrak{R}s = \mathrm{Ricci}(s^\#,.)$; in that case,

$$\Re_{min} = r_{min} \quad \text{(see nº 7.20);}$$

(ii) $E = \Lambda^p T^* M$, $\bar{D}$ is the Levi-civita connection, $\tilde{\Delta}$ is the Laplacian on p-forms, $\Re s$ can be expressed in terms of the curvature tensor of $(M,g)$; in that case $\Re_{min}$ can be computed in terms of upper and lower bounds on $\mathrm{Sect}(M,g)$ (see [LZ] p. 3, [G-M] p. 264 and [B-G] p. XV. 8);

(iii) $E = S(M)$, the bundle of spinors, $\bar{D}$ is the Levi-Civita connection, $\tilde{\Delta}$ is the Dirac operator, $\Re = \frac{u}{4}$ ($u$ is the scalar curvature of $(M,g)$: see nº III.45);

(iv) Other examples include the moduli space of Einstein metrics ([B-G] § 3), Jacobi fields for harmonic maps or minimal immersions ([E-L] or [UA]).

25. It follows from the Weitzenböck formula (22) that

$$\int_M \langle \tilde{\Delta} s | s \rangle dv_g = \int_M |\bar{D} s|^2 dv_g + \int_M \langle \Re s | s \rangle dv_g .$$

The following Proposition is a direct consequence of the above formula (same methods as in the proof of Proposition 20).

26. <u>Proposition</u>. <u>Under the above assumptions, let</u> $\delta(E)$ <u>denote the dimension of the space of harmonic sections of</u> $E$,

$$\delta(E) = \dim\{s \in C^\infty(E) \mid \tilde{\Delta} s = 0\}$$

(<u>this dimension is finite because</u> $\tilde{\Delta}$ <u>is elliptic</u>).

(i) <u>If</u> $\Re(x) \geq 0$, <u>and if there exists an</u> $x_o$ <u>in</u> $M$ <u>such that</u> $\Re(x_o) > 0$, <u>then</u> $\delta(E) = 0$.

(ii) <u>If</u> $\Re_{min} = 0$ <u>then</u> $\delta(E) \leq \ell = \mathrm{rank}(E)$.

Such results are called vanishing theorems. We will now deal with the following problem.

27. <u>Problem</u>. Give upper bounds on $\delta(E)$, in terms of estimates on $\Re(x)$ or $\Re_{min}$ and on the curvature of $(M,g)$.

Proposition 26 gives a partial answer to Problem 27 when $\Re_{min} \geq 0$. In the following paragraph, we investigate the case $\Re_{min} < 0$.

C. <u>THE ANALYTIC APPROACH, II</u>.

28. Assume that $\Re_{min} = -k^2$, $k \in \mathbb{R}_+^\bullet$, and let $s$ be a harmonic section of $E$, $\tilde{\Delta}s = 0$. It follows from nº 25 that

$$(29) \qquad \int_M |\bar{D}s|^2 v_g \leq k^2 \int_M |s|^2 v_g .$$

It follows from the very definition of $\bar{\Delta}$ and from the min-max principle (nº III.26), that (29) implies

$$\delta(E) \leq \text{number of eigenvalues of } \bar{\Delta} \text{ less than } k^2.$$

30. <u>Note</u>. The operators $\tilde{\Delta}$ and $\bar{\Delta}$ are non-negative, symmetric, $2^{nd}$ order elliptic linear partial differential operators on $C^\infty(E)$, so that the spectrum of $\tilde{\Delta}$ (resp. $\bar{\Delta}$) consists of a sequence of non-negative eigenvalues, with finite multiplicities

$$(0\leq)\ \tilde{\lambda}_1 \leq \tilde{\lambda}_2 \leq \dots \uparrow +\infty \quad (\text{resp. } (0\leq)\ \bar{\lambda}_1 \leq \bar{\lambda}_2 \leq \dots \uparrow +\infty).$$

In fact, the above inequality does not say much because we do not know the eigenvalues of $\bar{\Delta}$. Since $|s|$ is a function on $M$, we can try to obtain an inequality on scalar functions on $M$ (recall that we have bounds on the eigenvalues of $\Delta$ acting on $C^\infty(M)$ by Chapter V). For this purpose, we use the following lemma known as Kato's inequality (see [H-S-U]).

31. Lemma. <u>For any</u> $s$ <u>in</u> $C^\infty(E)$, <u>the pointwise norm of</u> $s, |s|$, <u>is in</u> $H^1(M,g)$ <u>and we have the following inequality (in the sense of distributions)</u>

$$|d|s|| \le |\bar{D}s|.$$

<u>Proof</u>. In the sense of distributions, we can write, for any $f$ in $C^\infty(M)$

$$\int_M (D_X|s|)f\, dv_g = -\int_M |s|\mathcal{L}_X(f\, dv_g)$$

$$= -\lim_{r\to 0}\int_M (|s|^2+r^2)^{1/2}\mathcal{L}_X(f\, dv_g)$$

$$= \lim_{r\to 0}\int_M \langle \bar{D}_X s|s\rangle(|s|^2+r^2)^{-1/2} f\, dv_g,$$

so that

$$d|s|(x) = \begin{cases} 0, & \text{if } s(x) = 0 \\ \langle \bar{D}_{\cdot} s|s\rangle_x / |s|_x, & \text{if } s(x) \ne 0, \end{cases}$$

and hence $d|s|$ is in $H^1(M)$ and satisfies

$$|d|s||_x \le |\bar{D}s|_x \quad \blacksquare$$

Note that $|d|s||$ is the norm of the 1-form $d|s|$ and that $|\bar{D}s|$ is the norm of the element $|\bar{D}s|$ in $T^*M\otimes E$.

In we apply Lemma 31 to (29), we find

$$\int_M |df|^2 dv_g \leq k^2 \int_M f^2\, dv_g, \quad \text{with } f = |s|. \tag{32}$$

From the min-max principle (nº III.26), we conclude that ... $k^2 \geq 0$, because $\lambda_1(M,g)$, closed) $= 0$ ... this is not very interesting! An interesting estimate would be $k^2 \geq \lambda_2(M,g)$, because we know how to bound $\lambda_2(M,g)$ from below by Cheeger's estimate (see nº IV. 26-29). In order to obtain such an estimate, we need to write (32) with a function $f$ such that $\int_M f = 0$ (see nº III. 26).

Define $h(x)$ by $h(x) = |s|_x - \int_M |s|_x\, dv_g \,/\, \mathrm{Vol}(M)$.

It is clear that $\int_M h = 0$. In order to substitute $h$ to $f$ in (32), we use the following general lemma due to Daniel Meyer.

33. Lemma ([ME2]). _Let_ $E$ _be a Riemannian vector bundle over the Riemannian manifold_ $(M,g)$. _Let_ $F$ _be a finite dimensional subspace of_ $L^2(E,v_g) = \{s \mid \int_M |s|_x^2\, dv_g(x) < \infty\}$, _such that_

$$\dim F = N > \ell = \mathrm{rank}(E).$$

_Then there exists an element_ $s_o$ _in_ $F$ _such that_

$$\mathrm{Vol}(M)^{-1/2} \int_M |s_o|_x dv_g \leq C(\ell,N)\left(\int_M |s_o|_x^2\, dv_g\right)^{1/2},$$

_where_ $C(\ell,N)$ _is a universal function of_ $(\ell,N)$ _which satisfies:_

$C(\ell,N)$ is strictly decreasing in $N$, $C(\ell,\ell) = 1$, $C(\ell,N)$ goes to $0$ when $N$ goes to infinity.

In order to apply Lemma 33, we take $F = \{s \in C^\infty(E) | \tilde{\Delta} s=0\}$. Then $N = \dim F = \delta(E)$. We assume that $N > \ell = \mathrm{rk}(E)$. Let $s_o$ be the section given by Lemma 33, and denote $|s_o|$ by $f$ and $f - \int_M f\, dv_g / \mathrm{Vol}(M)$ by $h$. We can then write

$$(1 - C^2(\ell,N)) \int_M f^2 \leq \int_M h^2,$$

so that (32) gives

$$\int_M |dh|^2 dv_g \leq k^2 (1-C^2(\ell,N))^{-1} \int_M h^2\, dv_g;$$

this last inequality implies that

$$(34) \qquad \lambda_2(M,g);\text{closed}) \leq k^2(1-C^2(\ell,N))^{-1}.$$

We can now prove the following generalization of Bochner's results.

35. Theorem. Let $E$ be a Riemannian vector bundle over the Riemannian manifold $(M,g)$ (compact without boundary) as in nº 21. Assume that

$$r_{min}(M,g)d(M,g)^2 \geq \epsilon(n-1)a^2,\ \epsilon \in \{-1,0,1\},\ a \in \mathbb{R}_+^{\bullet} \text{ and } n = \dim M$$

(see nº V. 26; the interesting case here is $\epsilon = -1$).

Then there exists a positive number $b = b(n,\epsilon,a)$ such that $\mathfrak{R}_{min} d(M,g)^2 \geq -b$ implies $\delta(E) \leq \ell$.

Proof. By Cheeger's inequality (nº IV. 26-29), there exists a constant $c = c(n,\varepsilon,a)$ such that

$\lambda_2(M,g;\text{closed}) \geq c/d(M,g)^2$. If $\delta(E) \geq \ell+1$ we can write, in view of (34) and Lemma 33 (see nº 28),

$$c(n,\varepsilon,a)(1-C^2(\ell,\ell+1)) \leq |\mathfrak{R}_{min}|\ d(M,g)^2.$$

This proves the theorem ■

36. <u>Example</u>. Take $E = T^*M$, $\delta(E) = b_1(M)$. We obtain

$$r_{min}(M)d^2(M) \geq -b(n,-1,1) \Rightarrow b_1(M) \leq n = \dim M,$$

which extends Bochner's result to the case in which the curvature of $(M,g)$ is allowed to take negative values.

37. <u>Comments</u>.

(i) Lemma 33 is quite general. It also applies to manifolds with boundary. See [ME2] for more applications;

(ii) Theorem 35 does not yet answer Problem 27. We could imagine to use (32) with enough functions $f$ in order to apply the variational characterization of eigenvalues (nº III. 28), and the estimates of Chapter V nº 31. The map $s \to |s|$ maps $C^\infty(E)$ to a <u>cone</u> in $H^1(M)$, so that it is not clear at all that one can apply the above idea (remember that if $s$ is a parallel section, then $|s|$ is a constant; in the case of a trivial bundle, the parallel sections form a vector space of dimension $\ell = rk(E)$, whose image by the application $s \to |s|$ is $\mathbb{R}_+$).

(iii) The first improvements of Bochner's result (i.e. when the curvature is allowed to take negative values) were given by P. Li (1980) for Betti numbers; they were then generalized by S. Gallot (1981). Both used Sobolev inequalities with Sobolev constants estimated in terms of isoperimetric inequalities. In 1980, M.Gromov gave bounds on the Betti numbers using geometric methods. He also pointed out that one should be able to use the heat equation and Kato's inequality. However he did not have the isoperimetric inequality for the heat kernel (see nº V. 28) and could therefore not go any further with this idea.

In the next paragraph we describe how heat kernel methods give partial answers to Problem 27.

## D. THE ANALYTIC APPROACH, III.

The idea is very simple. First of all notice that $\delta(E)$ is the multiplicity of 0 as eigenvalue of $\tilde{\Delta}$,

(38) $\delta(E) = \dim \operatorname{Ker} \tilde{\Delta}$.

It can be shown that the operator $\tilde{\Delta}$ (resp. $\bar{\Delta}$) has a heat kernel (see Chap. V § A), and that the trace of this heat kernel can be written as

$$\tilde{Z}(t) = \sum_{j=1}^{\infty} \exp(-\tilde{\lambda}_j t)$$

(resp. $\bar{Z}(t) = \sum_{j=1}^{\infty} \exp(-\bar{\lambda}_j t)$) (see nº 30),

where the series converges for $t > 0$.

Now recall that

$$\int_M \langle \tilde{\Delta} s | s \rangle \, dv_g \geq \int_M \langle \bar{\Delta} s | s \rangle \, dv_g + \mathfrak{R}_{min} \int_M \langle s | s \rangle \, dv_g,$$

so that the variational characterization of the eigenvalues (nº III.28) gives, for all $j \geq 1$, $\tilde{\lambda}_j \geq \bar{\lambda}_j + \Re_{min}$, from which we can deduce

(39) $\tilde{Z}(t) \leq \exp(-t\Re_{min})\bar{Z}(t)$, and hence

(40) $\delta(E) = \dim \operatorname{Ker} \tilde{\Delta} \leq \tilde{Z}(t) \leq \exp(-t\Re_{min})\bar{Z}(t)$, for all $t > 0$.

Finally we notice that

$$\operatorname{Ker} \bar{\Delta} = \{s \in C^\infty(E) \mid \bar{\Delta}s = 0\} \subset \{s \in C^\infty(E) | \bar{D}s = 0\}$$

so that

(41) $\dim \operatorname{Ker} \bar{\Delta} \leq \ell = \operatorname{rank}(E)$

(compare with Exercise 15 (ii)).

<u>Consequence</u>. Proposition 26 is an easy consequence of (39) - (41):

If $\Re_{min} > 0$, $\lim\limits_{t\infty} \tilde{Z}(t) = 0$, so that $\delta(E) = 0$;

If $\Re_{min} = 0$, $\lim\limits_{t\infty} \tilde{Z}(t) \leq \lim\limits_{t\infty} \bar{Z}(t) \leq \ell$, so that $\delta(E) \leq \ell$.

We now use the following theorem (see [H-S-U]), which generalizes Kato's inequality (Lemma 31).

42. <u>Theorem</u>. <u>Let</u> $E$ <u>be a Riemannian vector bundle of rank</u> $\ell$ <u>over the Riemannian manifold</u> $(M,g)$ (<u>see</u> nº 21). <u>Then</u>

$$\bar{Z}(t) \leq \ell \, Z(M,g;t), \quad \text{for all } t > 0.$$

43. <u>Remark</u>. Notice that equality holds in Theorem 42 when $E = M \times \mathbb{R}^\ell$.

Proof of Theorem 42. (for the results on operator theory we use here, see [KO], in particular Chap. 9).

Let $\varepsilon$ be a positive number. One can write (see Lemma 12)

$\frac{1}{2}\,\Delta(\langle s|s\rangle + \varepsilon^2) = \frac{1}{2}\,\Delta(\langle s|s\rangle) = \langle\bar{\Delta}s|s\rangle - |\bar{D}s|^2$. On the other hand,

$\frac{1}{2}\,\Delta(|s|_\varepsilon^2) = |s|_\varepsilon\,\Delta(|s|_\varepsilon) - |d|s_\varepsilon||^2$, where $|s|_\varepsilon^2 = \langle s|s\rangle + \varepsilon^2$,

so that

$$|s|_\varepsilon.\Delta(|s|_\varepsilon) = \langle\bar{\Delta}s|s\rangle + |d|s|_\varepsilon|^2 - |\bar{D}s|^2$$

from which we deduce (see Lemma 31),

$$|s|_\varepsilon.\Delta(|s|_\varepsilon) \le \langle\bar{\Delta}s|s\rangle.$$

Passing to the limit when $\varepsilon$ tends to $0$, this shows that

$$(*)\qquad |s|.\Delta(|s|) \le \langle\bar{\Delta}s|s\rangle,$$

where $\Delta(|s|)$ is understood in the sense of distributions, in particular this implies that $\Delta(|s|)$ is a measure (this extends Kato's inequality, Lemma 31).

For $\lambda \in \mathbb{R}_+^*$, we deduce from (*) that

$$|s|.(\Delta+\lambda\,\mathrm{Id})(|s|) \le \langle(\bar{\Delta}+\lambda\,\mathrm{Id})s|s\rangle.$$

Let $S = (\bar{\Delta}+\lambda\,\mathrm{Id})s$ so that, by Cauchy-Schwarz,

$$|(\bar{\Delta}+\lambda\,\mathrm{Id})^{-1}S|.(\Delta+\lambda\,\mathrm{Id})|(\bar{\Delta}+\lambda\,\mathrm{Id})^{-1}S| \le |S|\,|(\bar{\Delta}+\lambda\,\mathrm{Id})^{-1}S|$$

i.e.
$$(\Delta+\lambda\,\mathrm{Id})|(\bar{\Delta}+\lambda\,\mathrm{Id})^{-1}S| \le |S|.$$

Recall that $\exp(-t\Delta)$, the heat operator, preserves

positivity. For $\lambda > 0$, we write, in the sense of operators,

$$(\Delta+\lambda \mathrm{Id})^{-1} = \int_0^\infty e^{-t\lambda} e^{-t\Delta} dt,$$

so that $(\Delta+\lambda \mathrm{Id})^{-1}$ also preserves positivity. We conclude that for all $\lambda > 0$ and $n \in \mathbb{N}$,

$$|(\bar{\Delta}+\lambda \mathrm{Id})^{-n} S| \leq (\Delta+\lambda \mathrm{Id})^{-n}(|S|) \quad \text{and then}$$

$$\langle(\bar{\Delta}+\lambda \mathrm{Id})^{-n} S|T\rangle \leq |T|(\Delta+\lambda \mathrm{Id})^{-n}(|S|).$$

Recalling that $e^{-t\Delta} = \lim\limits_{t\infty} (1+\frac{t}{n}\Delta)^{-n}$, we conclude that for all $S,T$ in $C^\infty(E)$,

$$\langle \exp(-t\bar{\Delta}).S|T\rangle \leq |T| \exp(-t\Delta).(|S|)$$

and finally, we conclude that the norm $|\bar{k}(t,x,x)|$, of the endomorphism $\bar{k}(t,x,x)$ of the Euclidean space $E_x$, satisfies

$|\bar{k}(t,x,x)| \leq k_M(t,x,x)$, for any $x \in M$ ($\bar{k}(t,x,x)$ is the heat kernel of $\bar{\Delta}$).

We then deduce that $\bar{Z}(t) \leq \ell\, Z(t)$ ■ (Compare with Appendix A).

Finally the above results prove the following

44. Theorem. With the notations of nº 21, we have

$$\delta(E) \leq \ell \inf_{t>0} \exp(-t\mathfrak{R}_{min}) Z(M,g;t).$$

Theorem 44 together with Theorem V. 28 give partial answers to Problem 27.

45. Summary.

Let $(M,g)$ be a compact Riemannian manifold without boundary, such that

$$r_{min}(M,g)d(M,g)^2 \geq (n-1)\epsilon\alpha^2, \ \epsilon \in \{-1,0,1\}, \ \alpha \in \mathbb{R}_+^* \ \text{and } n=\dim M$$

(see nº V. 26).

Let $E \to M$ be a Riemannian vector bundle of rank $\ell$, equipped with a compatible connection $\bar{D}$, and a Laplacian $\tilde{\Delta}$ which satisfies the Weitzenböck formula

$$\tilde{\Delta}s = \bar{\Delta}s + \Re s .$$

Let $\Re_{min} = \inf\{\langle \Re s|s\rangle \mid s \in E, \langle s|s\rangle = 1\}$.

Then there exists a positive number $a(n,\epsilon,\alpha)$ (see nº V.26) such that

$$\delta(E) = \dim \operatorname{Ker} \tilde{\Delta} \leq \ell \inf_{t>0} F(t),$$

where $F(t) = \exp(-\Re_{min}d^2(M,g)t).Z(S^n,can;a^2(n,\epsilon,\alpha)t)$.

In particular, there exists a positive number $b(n,\epsilon,\alpha)$ such that $\Re_{min}d^2(M,g) \geq -b(n,\epsilon,\alpha)$ implies $\delta(E) \leq \ell$.

Note that since $Z(S^n,can;t)$ (see [CL] Chap. II. 4) and $a^2(n,\epsilon,\alpha)$ (see nº V. 26) are easily computable, the above estimate for $\delta(E)$ can be made very explicit, with intermediate values of $t$, i.e. with $t$ neither close to 0 nor very large. For explicit numerical computations, see [B-G] § 5.

46. Let $H_\gamma$ be a surface with genus $\gamma$ and constant curvature $-1$. The Gauss-Bonnet theorem ([HF] Part II Chap. III) implies that the volume of $H_\gamma$ is $4\pi(\gamma-1)$. Let $M_\gamma$ denote the Riemannian

product of $H_\gamma$ with a flat torus $T^{n-2}$ with volume $1/4\pi(\gamma-1)$. For $M_\gamma$, we have $r_{min}(M_\gamma) = -1$, $Vol(M_\gamma) = 1$. However $r_{min}(M_\gamma)d^2(M_\gamma)$ and $b_1(M_\gamma)$ tend to infinity with $\gamma$. This example shows that the above result (nº 45) is qualitatively best possible.

For more technical details, examples and counter-examples, we refer to [B-G] § 3; see also [GA2].

47. Remarks. In fact, Theorem V. 28 and Theorem 42 give the following estimates

$$\text{(i)} \quad Vol(M,g)k_M(t,x,x) \leq Z(S^n,can;a^2(n,\varepsilon,\alpha)t\ d(M)^{-2}),$$

$$\text{(ii)} \quad \tilde{k}_E(t,x,x) \leq \ell \exp(-t\mathfrak{R}_{min})\ k_M(t,x,x),$$

so that we also get bounds on the $L^\infty$-norms of the eigenfunctions of $\Delta$ on $C^\infty(M)$ or of the eigensections of $\tilde{\Delta}$ on $C^\infty(E)$.

## E. UNDERLYING PHILOSOPHY.

48. In nº IV. 12, we introduced the following class of Riemannian manifolds $\mathfrak{M}_{n,k,D} = \{(M,g) \mid \dim M = n,\ Ricci(M,g) \geq (n-1)kg,\ Diam(M,g) \leq D\}$.

In nº V. 26, we stated that $h(M,g;\beta)$ is bounded from below by a uniform function $H(\beta) = H(n,k,D;\beta)$ on $\mathfrak{M}_{n,k,D}$; in nº V. 28 we proved that $Z(M,g;t)$ is bounded from above by a uniform function $Z(t) = Z(n,k,D;t)$ on $\mathfrak{M}_{n,k,D}$. From the second estimate, we deduced, in this chapter, that $b_1(M)$ is uniformly bounded in the class of manifolds $M$ which admit a metric $g$ such that $(M,g)$ is in $\mathfrak{M}_{n,k,D}$. It is very important to visualize these results in the following picture.

49. Given two subspaces X and Y of a given metric space Z, we denote by $d_H^Z(X,Y)$ the infimum of the positive numbers r such that X (resp. Y) is contained in the r - neighborhood of Y (resp. X). This is the <u>Hausdorff distance</u> of X and Y in Z.

We now define the Hausdorff distance $d_H(X,Y)$ between two metric spaces X and Y as the infimum of the number $d_H^Z(i(X), j(Y))$, for all isometric embeddings $i: X \to Z$, $j: Y \to Z$ in the some metric space Z, as Z varies.

In [GV2], M. Gromov proves the following fundamental theorems

50. <u>Theorem</u>. (precompactness theorem: [GV2], Chap. 5). <u>The space $\mathfrak{M}_{n,k,D}$ is precompact for the Hausdorff distance $d_H$ between Riemannian manifolds (compact, without boundary)</u>.

Let $\mathfrak{M}_{n,K,D,V}$ denote the class of all Riemannian manifolds (M,g) such that $\dim M = n$, $|\mathrm{Sect}(M,g)| \leq K$, $\mathrm{Diam}(M,g) \leq D$ and $\mathrm{Vol}(M,g) \geq V$.

Let $\mathfrak{D}_{n,K,D,V}$ denote the class of all differentiable manifolds M such that there exists a metric g on M with $(M,g) \in \mathfrak{M}_{n,K,D,V}$.

51. <u>Theorem</u> (compactness theorem: [GV2], Chap. 8)

(i) (Cheeger) <u>The set</u> $\mathfrak{D}_{n,K,D,V}$ <u>is finite</u>;

(ii) (Gromov) <u>The set</u> $\mathfrak{M}_{n,K,D,V}$ <u>is compact for the Hausdorff distance</u> $d_H$ <u>and the map</u> $(M,g) \to M$ from $\mathfrak{M}_{n,K,D,V}$ <u>to</u> $\mathfrak{D}_{n,K,D,V}$ <u>is locally constant</u>.

It follows from Theorem 51 that $b_1(M)$, $b_p(M)$, and more generally any topological invariant is bounded on $\mathfrak{M}_{n,K,D,V}$. Indeed, such an invariant does not depend on the Riemannian metric, and

$\mathfrak{D}_{n,K,D,V}$ is finite[1], hence there are only finitely many homeomorphism classes of manifolds in $\mathfrak{D}_{n,K,D,V}$. Unfortunately, given $(n,K,D,V)$, we do not know in general the elements in $\mathfrak{D}_{n,K,D,V}$ so that we do not know any explicit bound on the topological invariants of the elements in $\mathfrak{M}_{n,K,D,V}$, at least from Theorem 51.

However, it follows from nº 45 that we can give explicit bounds on

(i) $b_1(M)$ in terms of $\dim(M)$, a lower bound on $\mathrm{Ricci}(M,g)$ and an upper bound on $\mathrm{Diam}(M,g)$;

(ii) $b_p(M)$, $2 \leq p \leq n-2$, in terms of $\dim(M)$, an upper bound on $|\mathrm{Sect}(M,g)|$ and an upper bound on $\mathrm{Diam}(M,g)$.

In view of (i) we could conjecture that any reasonable geometric invariant is bounded on $\mathfrak{M}_{n,k,D}$. Theorem 50 would prove this conjecture if these geometric invariants were continuous for the Hausdorff distance. This is not so in general, as the following examples show.

52. Counter-Examples. Let us consider $b_2(M)$, the second Betti number.

(i) $(M_\varepsilon,g) = (T^1,\mathrm{can}) \times (T^{n-1},\varepsilon\mathrm{can})$ converges to $(T^1,\mathrm{can})$ for the Hausdorff distance. However, $b_2(M_\varepsilon) = \binom{n}{2}$ and $b_2(T^1) = 0$.

(ii) Consider the Hopf fibration $S^{2n+1} \to \mathbb{C}P^n$, whose fiber is $S^1$. We can multiply the metric in the fiber by $\varepsilon$ so that we obtain a sequence $(S^{2n+1},g_\varepsilon)$ of manifolds whose $2^{nd}$ Betti number is $0$. This sequence converges for the Hausdorff distance to $(\mathbb{C}P^n,\mathrm{can})$,

---

(1) The number of elements in $\mathfrak{D}_{n,K,D,V}$ was recently bounded from above by S. Peters.

whose $2^{nd}$ Betti number is non-zero.

One can also show that $\lambda_2(M,g;closed)$ is not continuous for $d_H$ (see [TSG]).

53. Pre-compactness revisited.

Our Theorem V. 28 gives a precompactness result for $\mathfrak{M}_{n,k,D}$ which might have some relationship with Gromov's precompactness theorem (Theorem 50): see [B-B-G3].

We denote by $\ell^2$ the Hilbert space of sequences $\{a_i\}_{i\geq 1}$ such that $\Sigma|a_i|^2 < \infty$.

We denote by $h^1$ the Hilbert space of sequences $\{a_i\}_{i\geq 1}$ such that $\Sigma|a_i|^2(1+i^{2/n}) < \infty$ ($n$ fixed in $\mathbb{N}$).

It is a classical result that the inclusion $h^1 \to \ell^2$ is compact.

Now, given a Riemannian manifold $(M,g)$, we define a map $\psi$ from $(M,g)$ to $\ell^2$ as follows

$$\psi(x) = \{Vol(M,g)^{1/2} \exp(-\lambda_j)\phi_j(x)\}_{j\geq 1},$$

where $\{\lambda_j = \lambda_j(M,g;closed)\}_{j\geq 1}$ is the spectrum of the Laplacian on $C^\infty(M)$, and where $\{\phi_j\}_{j\geq 1}$ is an associated orthonormal family of eigenfunctions.

If $(M,g)$ is in $\mathfrak{M}_{n,k,D}$, we can write (use nº V. 28)

$$\text{(i)} \quad \|\psi(x)\|^2_{\ell^2} = Vol(M,g)k_M(2,x,x) \leq Z(S^n,can;A(n,k,D)), \quad \text{and}$$

$$\|\psi(x)\|^2_{h^1} = Vol(M,g) \sum_{j=1}^{\infty} (1+j^{2/n})\exp(-2\lambda_j)\phi_j^2(x).$$

Since $\lambda_j \geq B(n,k,D)j^{2/n}$ in view of Theorem V. 31, we conclude that

(ii) $\|\psi(x)\|^2_{h^1} \leq \mathrm{Vol}(M,g)C(n,k,D) \sum_{j=1}^{\infty} (1+\lambda_j)\exp(-2\lambda_j)\phi_j^2(x).$

Using nº V. 28 again and a summation by parts, we conclude that

(iii) $\|\psi(x)\|^2_{h^1} \leq E(n,k,D).$

Here $A(n,k,D)$, $B(n,k,D)$, $C(n,k,D)$, $E(n,k,D)$ are universal functions of $n,k,D$.

From (iii) we conclude that the image of the set $\mathfrak{M}_{n,k,D}$ by the application $\psi$ is bounded in $h^1$, and hence relatively compact in $\ell^2$. This is the announced precompactness result.

Further references for Chapter VI: For § B to D see also [BD1]. For § E, see [GV2] and [M-S], [SI].

# CHAPTER VII

## A BRIEF SURVEY OF SOME RECENT DEVELOPMENTS IN SPECTRAL GEOMETRY

ALL RIEMANNIAN MANIFOLDS ARE COMPACT,
CONNECTED, WITHOUT BOUNDARY

(Unless otherwise stated)

In Chapter III.C, we divided the problems concerning the relationship between the eigenvalues of the Laplacian and the geometry of a Riemannian manifold (spectral geometry) into two categories: direct problems and inverse problems. Both types of problems are relevant to mathematical physics.

In Chapters IV and V, we dealt with direct problems and more precisely with isoperimetric methods applied to direct problems (e.g. lower bounds on the eigenvalues). These chapters do not give an exhaustive survey of known results on direct problems. For more details we refer to [CL+BD2], [PE] and [ON]

In the present chapter, we give a brief overview of inverse problems. As we neither plan nor wish to give a thorough survey, we refer to [B-B] for references (see in particular the list of basic references given page 156 or Appendix B, p. 210).

In Chapter I, we motivated the study of eigenvalue problems by applying the method of separation of variables to the wave equation. We could have applied the same method to the heat equation. Now it turns out that this is the analysis of the heat and wave equations which leads to inverse results in spectral geometry.

## A. THE HEAT EQUATION AND APPLICATIONS.

1. In Chapter V § A, we introduced the heat kernel $k(t,x,y)$ of a Riemannian manifold $(M,g)$.

If we denote by $\{\lambda_i\}_{i\geq 1}$ the eigenvalues of the Laplacian $\Delta^g$ acting on $C^\infty(M)$ and by $\{\phi_i\}_{i\geq 1}$ an associated family of orthonormal real eigenfunctions, we can write (see nº V. 4)

$$\text{(i)} \quad k(t,x,x) = \sum_{j=1}^{\infty} \exp(-\lambda_j t)\phi_j^2(x),$$

$$\text{(ii)} \quad \int_M k(t,x,x)dv_g(x) = Z(t) = \sum_{j=1}^{\infty} \exp(-\lambda_j t).$$

Recall that $Z(t)$ determines $\{\lambda_i\}_{i\geq 1}$ (nº V. 5).

The following theorem (known as <u>Minakshisundaram - Pleijel asymptotic expansion</u>) has been used extensively to investigate inverse problems.

2. <u>Theorem</u>. <u>Let</u> $(M,g)$ <u>be an</u> n-<u>dimensional Riemannian manifold. The following asymptotic expansions hold when</u> $t$ <u>goes to</u> $0_+$

$$\text{(i)} \quad k(t,x,x) \sim (4\pi t)^{-n/2} \sum_{m=0}^{\infty} u_m(x)t^m;$$

$$\text{(ii)} \quad Z(t) \sim (4\pi t)^{-n/2} \sum_{m=0}^{\infty} a_m t^m;$$

(<u>these are asymptotic expansions; the series in the right hand sides do not converge in general</u>).

The functions $u_m(x)$ are $C^\infty$ functions on $M$ which can be expressed as universal polynomials in the components of the curvature tensor and its covariant derivatives.

In particular,

$$a_o = \mathrm{Vol}(M,g)$$

$$a_1 = \frac{1}{6}\int_M u(x)dv_g(x) \quad (u \text{ is the scalar curvature of } M\text{: II.45}).$$

3. Remarks. It is in fact very difficult to give explicit formulae for the functions $u_m$: see [B-G-M] Chap. III. E and [GY] or [BD].

In the sequel, we denote the sequence $\{\lambda_i\}_{i\geq 1}$ of the eigenvalues (with multiplicities) of the operator $\Delta^g$ acting on $C^\infty(M)$ by Spec(M,g) (the spectrum of (M,g)).

4. Some Consequences.

Assume we know the spectrum Spec(M,g). It follows from Theorem 2 that we then know

(i) the dimension of $M$,

(ii) the volume of $(M,g)$,

(iii) the integral $\int_M u(x)\,dv_g(x)$ of the scalar curvature of $(M,g)$, and hence, in dimension 2, the Euler-characteristic $\chi(M)$ of $M$, by the Gauss-Bonnet theorem.

5. Definition. We say that two Riemannian manifolds $(M,g)$ and $(N,h)$ are isospectral if Spec(M,g) = Spec(N,h).

One of the important questions in spectral geometry was formulated by M. Kac in the 1960's.

6. <u>Question</u>. "Can one hear the shape of a drum", or are two isospectral Riemannian manifolds isometric?

7. <u>Some positive answers to Question 6</u>.

(i) 2-dimensional flat tori are characterized by their spectra <u>among flat tori</u> ([B-G-M] Chap. III. B);

(ii) Let $(M,g)$ be one of the following Riemannian surfaces: $(S^2,\text{can})$, $(\mathbb{R}P^2,\text{can})$, $(T^2_\Gamma,\text{can})$. Then $(M,g)$ is characterized by its spectrum ([B-G-M] Chap. III. E);

(iii) $(S^n,\text{can})$ and $(\mathbb{R}P^n,\text{can})$ are characterized by their spectra for $n \leq 4$ ([B-G-M] Chap. III. E).

For further results see [B-B] Chap. 6.

8. <u>Some negative answers to Question 6</u>.

The first counter-example to Question 6 was given by J.Milnor in 1964. We can summarize the negative answers to Question 6 as follows

(i) There exist isospectral 16-dimensional flat tori which are not isometric (Milnor 1964); however, they are diffeomorphic;

(ii) There exist isospectral 5-dimensional lens spaces which are neither isometric nor homeomorphic (Ikeda 1980); (curvature+1);

(iii) There exist isospectral Riemann surfaces (with curvature-1) which are not isometric (they are homeomorphic by 4(iii)); there exist isospectral 3-dimensional manifolds with curvature-1, which are neither isometric nor homeomorphic (Vignéras 1980); recent examples were given by Sunada (1984) and Buser (1985);

(iv) There exists a <u>one-parameter family</u> of 5-dimensional Riemannian manifolds, such that any two elements of the family are isospectral but not isometric (C. Gordon - E. Wilson, 1983, J. Diff. Geom. 19 (1984)).

For further results see [B-B] Chap. 6 or references in § C.

9. Comments. The examples described by Milnor, Ikeda and Vignéras arise from number theoretic considerations; those of Gordon-Wilson from group theoretic considerations. It can be shown (Wolpert 1979) that there are finitely many non-isometric flat tori (resp. Riemann surfaces) with a given spectrum. However, no upper bound on the number of such tori (resp. Riemann surfaces) is known (except for 3-dimensional tori, J.P. Berry 1981).

In 1982, H. Urakawa gave the first examples of non-congruent domains in $\mathbb{R}^4$ with same Dirichlet and Neumann spectra. Examples of domains in $\mathbb{R}^3$ with the same property were recently given by P. Buser (1985). It would be interesting to have other counter-examples of such domains in $\mathbb{R}^n$ and specialy in $\mathbb{R}^2$.

More generally, it would be interesting to have examples of non-isometric, isospectral manifolds (possibly with boundary), which are not locally isometric.

We conclude this paragraph with two ramarks.

10. Remarks.

(i) The heat equation is a diffusion equation and is very much related to Brownian motion. Some results in spectral geometry are easily interpreted or proved in terms of Brownian motion and probability theory: see [CL] Chap. IX, [R-S] and [B-B] Chap. 12; probabilistic methods might turn out to be very powerful, for example to investigate the heat kernel with Dirichlet boundary condition in a domain with very irregular boundary.

(ii) A consequence of Theorem 2 (ii) is Weyl's asymptotic formula (see III. 36)

$$N(\lambda) = \mathrm{Card}\{j \mid \lambda_j \leq \lambda\} = C(n)\ \mathrm{Vol}(M,g)\lambda^{n/2} + o(\lambda^{n/2}), \tag{11}$$

which follows from the asymptotic formula for $Z(t)$ by applying Karamata's Tauberian theorem. However, one cannot give a sharp estimate for $N(\lambda) - C(n)\ \mathrm{Vol}(M,g)\lambda^{n/2}$ with this method. One has to use wave equation techniques.

## B. THE WAVE EQUATION AND APPLICATIONS.

The fundamental solution of the wave equation (or wave kernel) on $(M,g)$ is the distribution $E(t,x,y)$, $(t,x,y) \in \mathbb{R} \times M \times M$, which satisfies

$$\begin{cases} \left(\dfrac{\partial^2}{\partial t^2} + \Delta_y\right) E(t,x,y) = 0; \\ E(0,x,y) = \delta_x(y); \\ \dfrac{\partial E}{\partial t}(0,x,y) = 0. \end{cases} \tag{12}$$

In the sense of spectral theory, this is the kernel of the operator $\cos(t\sqrt{\Delta})$.

For example the wave kernel of $(\mathbb{R}^2, \mathrm{can})$ is

$$E(t,x,y) = C_2 |t| \left(|x-y|^2 - t^2\right)_-^{-3/2}, \quad \text{for some constant } C_2,$$

where

$$x_-^a = \begin{cases} 0 & \text{if } x \geq 0, \\ |x|^a & \text{if } x < 0. \end{cases}$$

The wave-kernel of $(T^2_\Gamma, can)$ is given by

(13) $$E(t,x,y) = C_2|t| \sum_{\gamma\in\Gamma} (\|x-y-\gamma\|^2 - t^2)_-^{-3/2}.$$

The wave equation techniques were introduced in the late 1960's by L. Hörmander to study the function $N(\lambda)$ (see nº 11). We now summarize the main results concerning the estimates on $N(\lambda)$.

14. Results on $N(\lambda)$. Let $(M,g)$ be a Riemannian manifold without boundary then

(i) $N(\lambda) = C(n)\mathrm{Vol}(M,g)\lambda^{n/2} + O(\lambda^{(n-1)/2})$ (Avakumovic 1956, Hörmander 1968) and this estimate is best possible, as the example of $(S^n, can)$ shows;

(ii) It was then observed by J. Duistermaat and V. Guillemin (1975) that the nature of $R(\lambda) = N(\lambda) - C(n)\mathrm{Vol}(M,g)\lambda^{n/2}$ is very much related to the geodesic flow of $(M,g)$. Roughly speaking, $R(\lambda)$ is of the order of $\lambda^{(n-1)/2}$ if and only if the geodesic flow of $(M,g)$ is periodic i.e. all geodesics are closed with same period. This is exactly why $(S^n, can)$ appears in (i). All the geodesics of $(S^n, can)$ are periodic with period $2\pi$. Again roughly speaking, if the geodesic flow is not periodic then $R(\lambda) = o(\lambda^{(n-1)/2})$;

(iii) In some cases, the estimate for $R(\lambda)$ can be improved. In the case of flat tori, one has $R(\lambda) = O(\lambda^{(n-2)/2+1/(n+1)})$. This estimate is not best possible, and the true nature of $R(\lambda)$ is not known; to investigate $R(\lambda)$ in that special case is a very difficult problem related to number theory. For manifolds with negative curvature one can prove that $R(\lambda) = O(\lambda^{(n-1)/2}/\mathrm{Log}\lambda)$ (Bérard, Randol 1976).

15. The case of manifolds with boundary is <u>much more difficult</u>. It can be shown on certain examples that the counting function $N_D(\lambda)$ for the Dirichlet eigenvalue problem in the manifold with boundary $(M,g)$ satisfies

$$(16) \quad N_D(\lambda) = C(n)\mathrm{Vol}(M,g)\lambda^{n/2} - C'(n)\mathrm{Vol}(\partial M,g)\lambda^{(n-1)/2} + o(\lambda^{(n-1)/2}).$$

This estimate is know as <u>Weyl's conjecture</u>.

Estimate (16) turns out to be much more difficult to prove than 14(i). In fact Hömander's estimate 14(i) was proved for manifolds with boundary only a few years ago (Pham The Lai, R.Seeley 1980). Counter-examples to Weyl's conjecture were given by D.Gromes (1967) and Bérard-Besson (1980): they again involve manifolds with boundary, whose geodesic flow (allowing reflections at the boundary, as in geometric optics) is periodic. Weyl's conjecture was <u>settled</u>, undergeneral assumptions, by Melrose (1980), Ivrii (1981) and Petkov (1985).

17. The wave kernel is very much related to the geodesic flow of $(M,g)$. In particular, wave equation techniques clarified the relationship between the spectrum of a Riemannian manifold and the lengths of the closed geodesics on the manifolds. Results in this direction were obtained by Y. Colin de Verdière (1973) and Duistermaat-Guillemin (1975). We explain this relationship for a flat torus $T^2_\Gamma$.

The Poisson summation formula (Example V. 6 (ii))

$$(4\pi t)^{-n/2}\mathrm{Vol}(T_\Gamma) \sum_{\gamma\in\Gamma} \exp(-\|\gamma\|^2/4t) = \sum_{\gamma^*\in\Gamma^*} \exp(-4\pi^2\|\gamma^*\|^2 t)$$

shows that the spectrum of $T_\Gamma$, $\{4\pi^2\|\gamma^*\|^2, \gamma^* \in \Gamma^*\}$ determines the lengths of the closed geodesics of $T_\Gamma$, $\{\|\gamma\|^2, \gamma \in \Gamma\}$, the so-called length-spectrum. This relation can also be seen as follows. Using nº (13) and the fact that $E(t,x,y)$ is the kernel of $\cos(t\sqrt{\Delta})$ we can write, at least at the formal level, for $T_\Gamma^2$

$$\sum_{j=1}^{\infty} \cos(t\sqrt{\lambda_j})\phi_j^2(x) = E_\Gamma(t,x,x) = C_2|t| \sum_{\gamma\in\Gamma} (\|\gamma\|^2-t^2)_-^{-3/2}.$$

In fact, one can show that $\sum_{j=1}^{\infty} \cos(t\sqrt{\lambda_j})$ is a tempered distribution whose singular support (points away from which the distribution is $C^\infty$) is contained in the set of lengths of closed geodesics (and their opposites). Again we see that the spectrum determines the set of lengths of the closed geodesics. This phenomenon can be understood by thinking of ripples propagating on a cylindrical lake: see Fig. 8 (think of a cylinder as a rectangle with two sides identified).

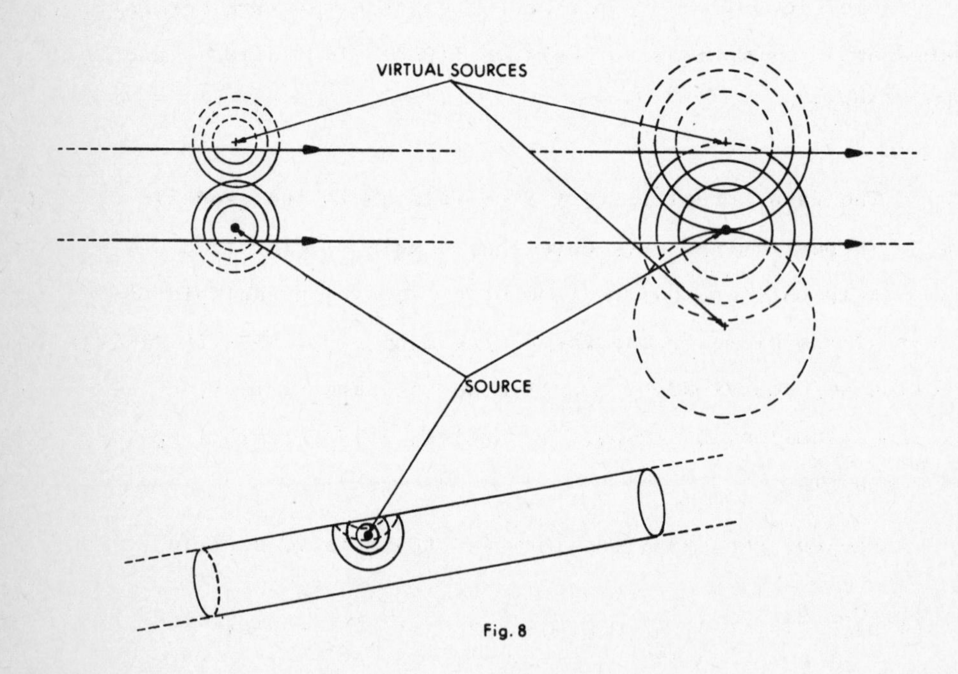

Fig. 8

Similar results can be established on manifolds with boundary, relating the spectra of the manifold, with either Dirichlet or Neumann boundary conditions, to billiard trajectories.

For more references on the wave equation see [B-B] Chap. E and § C. Let us end this paragraph by pointing out that wave equation techniques belong to the realm of symplectic geometry rather than to that of Riemannian geometry.

## C. FINAL COMMENTS.

Many problems arise in spectral geometry, both direct and inverse problems, both on manifolds with or without boundary. Leafing through the "Leitfaden" of [B-B], the reader will discover some of these problems. We only hope that these notes will arouse the interest of the readers and will lead them to solve some of these problems.

### Some further references

Heat equation: [BD], [CL+BD2], [DK]

Wave equation: [GL1], [GL2], [G-S]

Partial differential equations and geometry: [YU]

Open problems: [YU]

For the estimates concerning the counting functions $N(\lambda)$ (see (11)), we also refer the reader to: L. Hörmander, the analysis of partial differential operators, Vol.III-IV, Springer Grundlehren 1985.

General references: see [B-B] and the following review papers or books: [BE], [B-G-M], [BN], [CL+BD2], [GL1,2], [G-S], [GY], [PE], [P-S], [ON].

# APPENDIX A

## ON SYMMETRIZATION

by

GÉRARD BESSON[1]

(1) Institut Fourier, Math. Pures
Université de Grenoble
B.P. 74
38402 Saint Martin d'Hères

## I - INTRODUCTION

The spectral theory of Riemannian manifolds is a typical example of an interaction between two different aspects of mathematics: Riemannian geometry and operator theory in Hilbert spaces.

For geometers the aim is, of course, to obtain geometric informations using the well-known methods of Hilbert spaces analysis. This transfer can be summarized in the following picture:

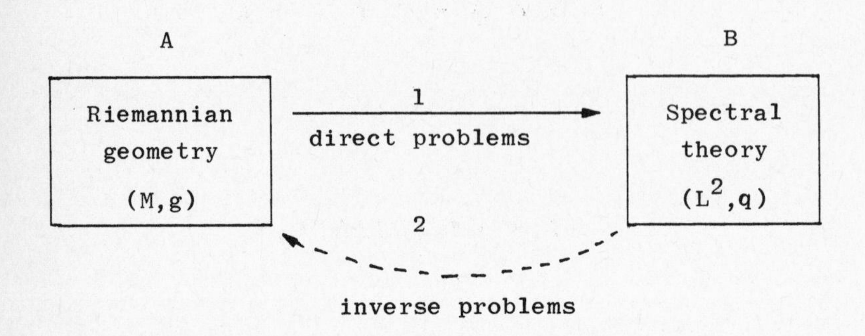


The link between box A and box B is done by associating to the metric space constituted by a Riemannian manifold $(M,g)$ another metric space built on the Hilbert space $L^2(M)$ endowed with a quadratic form $q$ in the sense of spectral theory (i.e. not necessarily everywhere defined) which we shall call the energy. In the case of a smooth compact connected Riemannian manifold, this quadratic form generates (by a standard process: the so-called Friedrichs' extension) a self-adjoint operator which has a compact resolvent. This is one of the best-known situation occuring in box B.

Once this link is established (may be the right words in this situation would be functors and categories, but such a formal approach is not our goal here) a lot of problems arise immediately. In Chapter III.C, the problems which are related to arrow 1 were called direct problems and those which come from interpretation of box B results, inverse problems. Let us now give some examples:

Question 1: Is arrow 1 injective?

This is the famous question asked by M. Kac [see Appendix B for references]: "Can one hear the shape of a drum?". It is known that the answer is no (in a general situation); however the following question has not yet been answered;

Question 2: Are two isospectral manifolds locally isometric?

Another interesting problem when dealing with the Laplacian (for example) would be to separate precisely PDE theoretic results from Hilbert spaces algebra ones, or in other words, to understand more accuratly the arrow 1.

This appendix illustrates an interesting principle which can be summarized as follows:

Once a problem in box A is translated into a problem in box B, it is very often useful to study thoroughly the latter problem before going back to the former one. This gives rise to more elementary and simple proofs.

The aim of the following text is to try to gather in a general picture a number of inequalities involving the heat semigroups. We wish to prove that such inequalities are (algebraic) consequences of inequalities on the quadratic forms associated with the operators under considerations, and are obtained in a manner which is very similar to the proof of Kato's inequality ([H-S-U1]: References p.190).

In the geometric applications it will then be clear that these criteria rely on isoperimetric inequalities. The former are in some sense, a Hilbert space translation of the latter. This relationship being established the desired results on diffusion processes are then easily proved.

Section II consists in recalling the Beurling-Deny criterion. We shall then give the corresponding criterion in our case, in box B, forgetting the geometrical meaning (Section IV). For this purpose we give a formal definition of the notion of symmetrization. The applications (Section V, VI) then consist in a verification of the conditions established in Section IV (geometric symmetrization decreases energies) in the particular cases under considerations: Fiber bundles, Schwarz symmetrization. Recall that a symmetrization generally yields a comparison between a generic space and a more symmetric one, in the sense that it has more isometries. Certain results are then incorporated in this framework such as the paper [B-G] which has been a guide for this text.

The reader will easily see that this appendix deals neither with the most general case nor with the most formal one, and that extensions and modifications are possible.

Finally, it must be noted that the idea of the formal approach of the inequalities appearing in [B-G] has been suggested to the author by the article [H-S-U1]. Our method is just a slight modification of the one presented in [H-S-U1]. In a forthcoming paper we shall develop a similar approach to the notion of transplantation (which is in some sense dual to that of symmetrization).

The results which we present in this Appendix are not yet polished, as they pertain to current research. A hopefully more satisfactory version of this Appendix will appear elsewhere.

References for this Appendix are given p. 190.

## II - THE ABSOLUTE VALUE AS A SYMMETRIZATION PROCEDURE AND THE BEURLING-DENY CRITERION

This section is our first contact with symmetrization, in its weakest form. However all the ideas (which are very simple) will appear here.

Let $(M,\mu)$ be a $\sigma$-finite measure space. We will deal with the Hilbert space $L^2(M,d\mu)$. It is then natural to give the

1. Definition:

*Let* $A$ *be a bounded operator on* $L^2$, *it is called positivity preserving if* $Af$ *is positive whenever* $f$ *is positive* (see [R-S]IV, p. 201).

We can now give sufficient conditions for a self-adjoint operator to be positivity preserving.

2. Proposition:

*Let* $H$ *be a self-adjoint operator, bounded from below by* $E = \inf\{\mathrm{Spec}(H)\}$. *Then* $e^{-tH}$ *is positivity preserving for all* $t > 0$ *if and only if* $(H-\lambda)^{-1}$ *is positivity preserving for all* $\lambda < E$.

Proof. Use the formulae

$$(H-\lambda)^{-1} = \int_0^\infty e^{\lambda t}\, e^{-tH} dt \qquad (\lambda < E)$$

$$e^{-tH} = \lim_{n\to+\infty} \left(1+\frac{tH}{n}\right)^{-n} \qquad (t > 0). \blacksquare$$

In the following we will only consider self-adjoint operators and real valued functions (for the sake of simplicity only).

There is a very simple criterion for a positive self-adjoint operator to generate a positivity preserving semi-group.

3. Theorem (Beurling-Deny criterion)

*Let* $H \geq 0$ *be a self-adjoint operator on* $L^2(M,d\mu)$ *and let* $q$ *be the associated quadratic form with domain* $\mathcal{D}(q)$. *The following assertions are equivalent*:

a) $e^{-tH}$ *is positivity preserving for all* $t > 0$,

b) *if* $u \in \mathcal{D}(q)$ *then* $|u| \in \mathcal{D}(q)$ *and* $q(|u|) \leq q(u)$.

Remark.

As in Proposition 2, it is sufficient to have $H$ bounded from below by some constant $-c$, i.e.

$$\langle H\varphi|\varphi\rangle \geq -c|\varphi|^2, \quad \text{for all } \varphi \text{ in the domain of } H.$$

Indeed, in that case $H + c\mathrm{Id}$ is non-negative, and the quadratic form associated to this operator is,

$$q(\varphi) + c|\varphi|^2.$$

The proof is very easy and it can be found in [R-S]IV,p.210.

In 1973, T. Kato [KO] proved a simple but very useful inequality, the so-called "Kato's inequality" (nº VI.31 and proof of VI.42). Later B. Simon [SN] gave an interpretation of this inequality in terms of positivity preserving semigroup. More precisely,

4. Definition:

*With the notations of the above theorem, we say that* $H$ *obeys Kato's inequality if and only if:*

i) $u \in \mathcal{D}(q)$ *implies* $|u| \in \mathcal{D}(q)$

ii) *for* $u \in \mathcal{D}(H)$ *and* $f \in \mathcal{D}(q)$ *with* $f \geq 0$

$$\langle f | H(|u|) \rangle \leq \langle (\text{sign } u) f | Hu \rangle$$

*where* $\text{sign}(u)$ *is defined to be*

$$(\text{sign } u)(x) = 0 \quad \textit{if} \quad u(x) = 0$$

$$= \frac{u(x)}{|u(x)|} \quad \textit{otherwise.}$$

The link with positivity is given by the following

5. Theorem [SN]

*A self-adjoint operator* $H$ *which is bounded from below satisfies Kato's inequality if and only if* $e^{-tH}$ *is positivity preserving for all* $t$.

*Proof*. If Kato's inequality holds, taking $f = |u|$ we get, for $u \in \mathcal{D}(H)$

$$q(|u|) \leq q(u).$$

A limiting argument and the use of Beurling-Deny criterion give the result.

Conversely if $e^{-tH}$ is positivity preserving, then for any $u$ and any $f \geq 0$

$$\langle (\text{sign } u) f | e^{-tH} u \rangle \leq \langle f | e^{-tH} |u| \rangle$$

and equality holds at $t = 0$.

If $u \in \mathcal{D}(H)$, taking $f = |u|$ and differentiating at $t = 0$ gives

$$0 \le q(|u|) \le q(u) \text{ and hence,}$$

$|u| \in \mathcal{D}(q)$. A limiting argument shows that the same is true when $u \in \mathcal{D}(q)$. If $u \in \mathcal{D}(H)$ and $f \in \mathcal{D}(q)$, differentiating again gives the desired inequality. ∎

6. <u>Interpretation</u>:

Let us assume that the self-adjoint operator $H$ is bounded from below and satisfies the Beurling-Deny criterion. Then the absolute value can be seen of as a mapping from $L^2(M,d\mu)$ to itself which decreases the quadratic form associated to $H$ (and thus which preserves its domain). Assume that $H \geq -c\, \mathrm{Id}$.

The consequences are the positivity preserving properties of $(H+\lambda)^{-1}$ (for $\lambda > c$) and $e^{-tH}$ (for $t > 0$), and Kato's inequality.

If $(H+\lambda)^{-1}$ (resp. $e^{-tH}$) is an integral operator with kernel $R(\lambda;.,.)$ (resp. $K(t;.,.)$), this leads to the positivity of the function on $M \times M$, $R(\lambda;.,.)$, for $\lambda > c$ (resp. $K(t;.,.)$ for $t > 0$).

So this property of the absolute value allows us to compare the operator to $0$, the trivial operator (positivity). The question which arises now is:

<u>Is it possible to compare different operators, even acting on different Hilbert spaces?</u>

This section is devoted to a formal approach of symmetrization. It aims at giving a criterion analogous to that of Beurling-Deny for semigroup domination. It is clearly inspired by the paper [H-S-U1].

Unless otherwise specified we shall deal with self-adjoint operators. This is not really necessary.

Let $\mathcal{H}$ and $\mathcal{K}$ be two Hilbert spaces. We assume $\mathcal{K}$ to be real.

A nonempty cone $\mathcal{K}^+$ of $\mathcal{K}$ is a subset such that:

i) $\mathcal{K}^+ + \mathcal{K}^+ \subseteq \mathcal{K}^+$,

ii) $a\mathcal{K}^+ \subseteq \mathcal{K}^+$ for all $a \geq 0$,

we assume furthermore that

iii) $\langle \mathcal{K}^+ | \mathcal{K}^+ \rangle \geq 0$.

The cone will be said to be <u>self-dual</u> if the following condition holds

iv) $\langle g | \mathcal{K}^+ \rangle \geq 0$ implies $g \in \mathcal{K}^+$.

In this situation, we have the

7. <u>Definition</u>:

<u>A map</u> $S$ <u>from</u> $\mathcal{H}$ <u>to</u> $\mathcal{K}^+$ <u>is called a symmetrization if</u>

1) <u>for all</u> $(f,f')$ <u>in</u> $\mathcal{H}$, $|\langle f|f'\rangle| \leq \langle S(f)|S(f')\rangle$, <u>and equality if</u> $f = f'$;

2) <u>Let</u> $g$ <u>be any element in</u> $\mathcal{K}^+$, <u>then for any</u> $f_1 \in \mathcal{H}$ <u>there exists an</u> $f_2 \in \mathcal{H}$ <u>such that</u>

$$g = S(f_2) \quad \underline{\text{and}}$$

$$\langle f_1|f_2\rangle = \langle S(f_1)|S(f_2)\rangle = \langle S(f_1)|g\rangle$$

(*in this case* $f_1$, $f_2$ *are said to be* g-*paired*).

We then have the

8. *Proposition*:

*A symmetrization is a Lipschitz map.*

*Proof*. $|S(f)-S(g)|^2 = |S(f)|^2 + |S(g)|^2 - 2\langle S(f)|S(g)\rangle$

but $|S(f)|^2 = |f|^2$ and

$$\langle f|g\rangle \leq |\langle f|g\rangle| \leq \langle S(f)|S(g)\rangle,$$

then

$$|S(f)-S(g)|^2 \leq |f-g|^2. \blacksquare$$

9. *Remark*.

This property allows us to define the symmetrization on a dense subset of $\mathcal{H}$ only.

Another property which will be important in the sequel is given by the following:

10. *Proposition*:

*Let* $\mathcal{H},\mathcal{K}$ *and* $\mathcal{L}$ *be three Hilbert spaces*, $\mathcal{K}$ *and* $\mathcal{L}$ *being real*, S (*resp*. T) *a symmetrization map from* $\mathcal{H}$ *to* $\mathcal{K}^+$ (*resp. from* $\mathcal{K}$ *to* $\mathcal{L}^+$); *if* T *has the property that whenever* $f_1 \in \mathcal{K}^+$ *and* $g \in \mathcal{L}^+$, *the element* $f_2 \in \mathcal{K}$ *such that* $(f_1,f_2)$ *are* g-*paired can be chosen in* $\mathcal{K}^+$ *then* $T\circ S$ *is a symmetrization*.

*Proof*. Clear.

The following proposition will lead to a definition of the domination relation.

11. Proposition:

*Let* $A$ *(resp.* $B$*) be a bounded operator on* $\mathcal{H}$ *(resp.* $\mathcal{K}$*). The following inequalities are equivalent (for all* $f_1, f_2 \in \mathcal{H}$ *and* $g \in \mathcal{K}^+$*):*

i) $\langle S(Af_1)|g\rangle \leq \langle B(S(f_1))|g\rangle$;

ii) $\Re e\langle Af_1|f_2\rangle \leq \langle B(S(f_1))|S(f_2)\rangle$;

iii) $|\langle Af_1|f_2\rangle| \leq \langle B(S(f_1))|S(f_2)\rangle$;

*if futhermore* $\mathcal{K}^+$ *is a self-dual cone, we can add*

iv) $S(Af_1) \leq B(S(f_1))$, i.e. $B(S(f_1)) - S(Af_1) \in \mathcal{K}^+$.

Proof of the Proposition:

(iii) $\Rightarrow$ (ii) trivial

(ii) $\Rightarrow$ (i)

Choose $f_2$ such that $(Af_1, f_2)$ are g-paired, then

$$\langle S(Af_1)|g\rangle = \langle Af_1|f_2\rangle = \Re e\langle Af_1|f_2\rangle \leq \langle B(S(f_1))|g\rangle.$$

(i) $\Rightarrow$ (iii) clear.

Finally it is clear that (iv) implies (i), (ii) and (iii), and that the self-dual property of $\mathcal{K}^+$ allows the converse to be true. ■

12. Definition:

*If* $A$ *and* $B$ *satisfy one of these inequalities we will say that* $B$ *dominates* $A$.

13. Remark.

The fact that the cone is self-dual allows to pass from integral inequalities to pointwise ones when working on spaces of functions. This will be important in the applications and explains the differences between inequalities obtained from various types of symmetrizations.

The following lemma will be important in the sequel.

14. Lemma:

1) If $B_i$ dominates $A_i$ then $|\alpha_1|B_1 + |\alpha_2|B_2$ dominates $\alpha_1 A_1 + \alpha_2 A_2$ $(\alpha_1, \alpha_2 \in \mathbb{C})$.

2) If $B_i$ dominates $A_i$ and if $B_i$ and $A_i$ converge respectively to $B$ and $A$ (weakly or strongly) then $B$ dominates $A$.

3) If $B_i$ dominates $A_i$ $(i=1,2)$ and if $B_1$ preserves the cone $\mathcal{K}^+$ then $B_1 \circ B_2$ dominates $A_1 \circ A_2$. If furthermore $\mathcal{K}^+$ is self-dual then the preservation of $\mathcal{K}^+$ by $B_1$ is a consequence of the domination relation.

Proof. The only non trivial point is 3). For $f \in \mathcal{H}$, $g \in \mathcal{K}^+$ we have

$$\langle S(A_1 \circ A_2(f)) | g \rangle \leq \langle B_1(S(A_2 f)) | g \rangle \leq \langle S(A_2 f) | B_1 g \rangle$$

(recall that the operators are assumed to be self-adjoint). Then

$$\langle S(A_1 \circ A_2(f)) | g \rangle \leq \langle B_2(S(f)) | B_1 g \rangle = \langle (B_1 \circ B_2)(S(f)) | g \rangle,$$

because $B_1 g$ belongs to $\mathcal{K}^+$.

If $\mathcal{K}^+$ is self-dual then the relation

$$S(A_1 f) \leq B_1(S(f))$$

implies that for all $g \in \mathcal{K}^+$, choosing $f \in \mathcal{H}$ such that $S(f) = g$, we have

$$\langle B_1 g | h \rangle \geq \langle S(A_1 f) | h \rangle \geq 0 \quad \text{for all} \quad h \in \mathcal{K}^+.$$

Thus

$$B_1 g \in \mathcal{K}^+. \quad \blacksquare$$

15. <u>Corollary</u>:

<u>Let</u> $H$ (<u>resp</u>. $K$) <u>be a self-adjoint operator on</u> $\mathcal{H}$ (<u>resp</u>. $\mathcal{K}$) <u>both bounded from below by</u> $-c$ <u>and let</u> $P_t = \exp(-tH)$ <u>and</u> $T_t = \exp(-tK)$. <u>The following propositions are equivalent</u>:

i) $T_t$ <u>dominates</u> $P_t$ <u>for all</u> $t > 0$;

ii) $(\lambda+K)^{-1}$ <u>dominates</u> $(\lambda+H)^{-1}$ <u>for all</u> $\lambda > c$;

iii) $(\lambda+K)^{-n}T_t$ <u>dominates</u> $(\lambda+H)^{-n}P_t$ <u>for all</u> $\lambda > c$, $t > 0$ <u>and</u> $n \in \mathbb{N}$.

<u>Proof</u>. Use the formulae

$$(\lambda+H)^{-1} = \int_0^{+\infty} e^{-t\lambda} P_t \, dt \qquad (\lambda > c),$$

$$P_t = \lim_{n\to+\infty} (1+ \frac{t}{n} H)^{-n} = \lim_{n\to+\infty} [(\frac{n}{t})^n(\frac{n}{t} + H)^{-n}] \quad \blacksquare$$

## IV - INTERPRETATION IN TERMS OF QUADRATIC FORMS: A CRITERION FOR SEMIGROUP DOMINATION

We can now give a necessary and sufficient condition in terms of quadratic forms for a symmetrization to give rise to semigroup domination. It will be a generalization of Beurling-Deny's criteria.

The main theorem of this text is:

16. Theorem:

Let $H$ and $K$ be self-adjoint operators, both bounded from below by $-c$, on $\mathcal{H}$ and $\mathcal{K}$ respectively; $q_H$, $q_K$ the associated quadratic forms (which will be considered as bilinear forms as well). Let $P_t$ and $T_t$ be the semigroups generated by $H$ and $K$. If $\mathcal{D}_0$ is a core for $H$, and if we assume that $(\lambda+K)^{-1}$ preserves $\mathcal{K}^+$ for all $\lambda > c$, then the following conditions are equivalent:

a) Semigroup domination

$T_t$ dominates $P_t$ for all $t > 0$;

b) Resolvent domination

$(\lambda+K)^{-1}$ dominates $(\lambda+H)^{-1}$ for all $\lambda > c$;

c) Kato's inequality

$(K_1)$ $u \in \mathcal{D}(q_H)$ implies $S(u) \in \mathcal{D}(q_K)$

$(K_2)$ $|q_H(f_1,f_2)| = |\langle Hf_1|f_2\rangle| \geq \mathcal{R}e\langle Hf_1|f_2\rangle \geq$

$$\geq \langle S(f_1)|KS(f_2)\rangle = q_K(S(f_1),S(f_2)),$$

for all $f_1 \in \mathcal{D}_0$, $f_2$ such that $S(f_2) \in \mathcal{D}(K)$ and $(f_1,f_2)$ are $S(f_2)$-paired.

Proof.

i) The equivalence of a) and b) is clear from the previous section.

ii) a) implies c). The hypothesis a) implies the inequality:

$$\langle (\frac{1-e^{-tH}}{t}) f | f \rangle \geq \langle (\frac{1-e^{-tK}}{t}) S(f) | S(f) \rangle ,$$

for $f \in \mathcal{D}_0$. Letting $t$ go to zero yields,

$$+\infty > q_H(f) = |H^{1/2} f|^2 \geq |K^{1/2} S(f)|^2 = q_K(S(f)) \geq 0,$$

which implies

$$S(f) \in \mathcal{D}(q_K).$$

A similar argument gives inequality $(K_2)$.

Recall that a core for $H$ is a subset $\mathcal{D}_o$ of $\mathcal{D}(H)$ such that the graph of $H|_{\mathcal{D}_o}$ is dense in the graph of $H$.

iii) c) implies b).

Let us assume that $f_1 \in (\lambda+H)(\mathcal{D}_o)$ and $g \in \mathcal{H}^+$, then

$$\langle S((H+\lambda)^{-1} f_1) | g \rangle = \langle S(h_1) | g \rangle ,$$

with $h_1 = (H+\lambda)^{-1} f_1$.

Since $h_1 \in \mathcal{D}_o$, and by assumption $(\lambda+K)^{-1} g \in \mathcal{H}^+$, we can write

$(\lambda+K)^{-1} g = S(f_2)$ with $h_1$ and $f_2$, $S(f_2)$-paired.

(Notice that $S(f_2) \in \mathcal{D}(K)$). Then (by (K2))

$$\langle S(h_1) | g \rangle = \langle S(h_1) | (\lambda+K) S(f_2) \rangle \leq \Re e \langle (H+\lambda) h_1 | f_2 \rangle = \Re e \langle f_1 | f_2 \rangle \leq$$

$$\leq |\langle f_1 | f_2 \rangle| \leq \langle S(f_1) | S(f_2) \rangle = \langle S(f_1) | (\lambda+K)^{-1} g \rangle .$$

Then we have

$$\langle S((H+\lambda)^{-1}f_1)|g\rangle \leq \langle(\lambda+K)^{-1}S(f_1)|g\rangle,$$

which is the desired inequality. A limiting argument allows us to conclude. ■

Let $Eg(H)$ (resp. $Eg(K)$) be the ground state energy of the self-adjoint operator $H$ (resp. $K$), i.e. the infimum of the spectrum of $H$ (resp. $K$), the following consequence is immediate.

17. Corollary:

If one of the conditions in Theorem 16 is verified then,

$$Eg(H) \geq Eg(K).$$

Proof. Use the min-max principle. ■

18. Interpretation:

The theorem can be summarized in:

the semigroup $P_t$ is dominated by $T_t$ if and only if $S$ does not increase the energy integral.

19. Remarks.

i) We assumed the operators to be self-adjoint. Clearly, this class can be enlarged. As an example, in [H-S-U1], the operators are just assumed to be maximally accretive.

ii) If $e^{-tH_i}$ is dominated by $e^{-tK_i}$ $(i=1,2)$ and if $H_1+H_2$ and $K_1+K_2$ are in the class of operator under consideration then $e^{-tH}$ is dominated by $e^{-tK}$, $H$ and $K$ being the closures of $H_1+H_2$ and $K_1+K_2$ respectively. This is easily proved by applying Trotter product Formula.

iii) As a conclusion, let us summarize the ideas of the three preceding sections in the following comparative chart:

| <u>Beurling-Deny-Simon Criterion</u> | <u>Generalized Criterion</u> |
|---|---|
| $H$ self-adjoint operator in $L^2(M,d\mu)$ bounded from below by $-c$. $q_H$ associated quadratic form. | $H$ self-adjoint operator on $\mathcal{H}$, $H \geq -c$; $q_H$.<br>$K$ self-adjoint operator on $\mathcal{K}$, $K \geq -c$; $q_K$.<br>$S$ symmetrization from $\mathcal{H}$ to $\mathcal{K}$ with positive cone $\mathcal{K}^+$ preserved by $(\lambda+K)^{-1}$ for all $\lambda > c$.<br>$\mathcal{D}_o$ a core for $H$. |
| The following assertions are equivalent | |
| i) $e^{-tH}$ positivity preserving for all $t > 0$;<br>ii) $(H+\lambda)^{-1}$ positivity preserving for all $\lambda > c$;<br>iii) $u \in \mathcal{D}(q_H) \Rightarrow \lvert u\rvert \in \mathcal{D}(q_H)$ and $q_H(\lvert u\rvert) \leq q_H(u)$;<br>iv) $u \in \mathcal{D}(q_H) \Rightarrow \lvert u\rvert \in \mathcal{D}(q_H)$ and $\forall u \in \mathcal{D}(H)$ $\forall f \in \mathcal{D}(q_H)$, $f \geq 0$ $\langle f \mid H(\lvert u\rvert)\rangle \leq \langle \mathrm{sign}(u) f \mid H(u)\rangle$ | i) $e^{-tK}$ dominates $e^{-tH}$ for all $t > 0$;<br>ii) $(\lambda+K)^{-1}$ dominates $(\lambda+H)^{-1}$ for all $\lambda > c$;<br>iii) Kato's inequality<br>$(K_1)$ $u \in \mathcal{D}(q_H) \Rightarrow S(u) \in \mathcal{D}(q_K)$<br>$(K_2)$ for all $f_1 \in \mathcal{D}_o$ and for all $f_2$ such that $S(f_2) \in \mathcal{D}(K)$ and $(f_1,f_2)$ are $S(f_2)$-paired, $\mathrm{Re}\langle Hf_1 \mid f_2\rangle \geq q_K(S(f_1),S(f_2))$. |

Kato's inequality is the subject of the article [H-S-U1]. For an application of this inequality, see [H-S-U] or Chap. VI. §III. Let us briefly summarize the situation.

Let $M$ be a compact Riemannian manifold and $E \xrightarrow{\pi} M$ a hermitian vector bundle, i.e. a vector bundle such that each fiber is equipped with a hermitian structure varying smoothly in the base point.

Let $D$ be a connection on the sections of $E$, compatible with the hermitian product.

The Riemannian metric on $M$ gives rise to the Laplace-Beltrami operator $\Delta$.

The connection $D$ together with the Riemannian structure on $M$ allow to define a Laplacian type operator on the space $L^2(M;E)$ of $L^2$-sections of the bundle $E$, called the <u>rough Laplacian</u> $\bar{\Delta}$, and which is a non-negative self-adjoint operator (see VI.13).

Finally, for a section $u$ in $L^2(M;E)$, we define the function $|u|$ in $L^2(M;\mathbb{R})$ by the relation,

$$|u|(m) = |u(m)| \quad \text{for all} \quad m \in M.$$

The norm in the right hand side is taken in the fiber $E_m$.

We then have the

20. <u>Theorem</u> ([H-S-U1])

<u>With the above notations</u>

$$|e^{-t\bar{\Delta}}(u)| \leq e^{-t\Delta}(|u|).$$

Proof. The proof is done by showing that the map

$$S: L^2(M;E) \to L^2(M,\mathbb{R})$$

$$u \to |u|$$

is a symmetrization which decreases energies. Recall that the quadratic form defining $\bar{\Delta}$ is

$$\bar{q}(u) = \int_M |Du|^2.$$

i) Define, for $\varepsilon > 0$ and $s$ a smooth section of the bundle

$$|s|_\varepsilon = (|s|^2 + \varepsilon^2)^{1/2}$$

which is then a smooth function on the manifold $M$. An easy computation gives

$$\Delta|s|_\varepsilon = \frac{(\bar{\Delta}s,s)}{|s|_\varepsilon} - \left(\frac{|Ds|^2}{|s|_\varepsilon} - \frac{|(Ds,s)|^2}{|s|_\varepsilon^3}\right).$$

Here $Ds$ is considered as a one-form with values in the bundle and $(Ds,s)$ a real one-form, and $(.,.)$ is the scalar product in each fiber of the bundle. By the Cauchy-Schwarz inequality

$$\frac{|(Ds,s)|^2}{|s|_\varepsilon^3} \leq \frac{|Ds|^2|s|^2}{|s|_\varepsilon^3} \leq \frac{|Ds|^2}{|s|_\varepsilon} .$$

Therefore,

$$\Delta|s|_\varepsilon \leq (\bar{\Delta}s, \frac{s}{|s|_\varepsilon}) .$$

ii) Let $\mathcal{D}_o$ be the space of $C^\infty$ sections. It is a core for $\bar{\Delta}$. Let $s_1$ be in $\mathcal{D}_o$ and $g$ a non-negative function on $M$ in the domain of $\Delta$. Define

$$
s_2(m) = \begin{cases} g(m)\,\dfrac{s_1}{|s_1|}(m) & \text{if} \quad s_1(m) \neq 0 \\ \\ g(m)e(m) & \text{if} \quad s_1(m) = 0 \end{cases}
$$

where $e(m)$ is any fixed measurable section of $E$ over $M$ such that

$$
|e(m)| = 1 \quad \text{for all} \quad m.
$$

(such sections exist since away from a set of measure zero the manifold is contractible).

Then clearly $s_2$ and $s_1$ are g-paired.

Furthermore,

$$
\langle |s_1|_\epsilon \big| \Delta g\rangle = \langle \Delta |s_1|_\epsilon \big| g\rangle = \int_M (\Delta |s_1|_\epsilon) g \leq
$$

$$
\leq \int_M (\bar{\Delta} s_1, g\, \frac{s_1}{|s_1|_\epsilon}) = \langle \bar{\Delta} s_1 \big| g\, \frac{s_1}{|s_1|_\epsilon}\rangle
$$

since $g$ is non-negative and in the domain of $\Delta$.

Letting $\epsilon$ go to zero yields

$$
\langle \bar{\Delta} s_1 | s_2\rangle \geq \langle |s_1| \big| \Delta |s_1|\rangle
$$

then the result follows by Theorem 16. ■

Let us point out that in that case the cone is self-dual since it is the set of non-negative functions of $L^2(M;\mathbb{R})$ and thus allows to get a strong domination inequality in Theorem 20.

This is the key section of this appendix. We aim at giving an alternative proof of Theorems V.9 and V.28, using the formal approach of symmetrization. For the sake of simplicity, we will give the construction in the simpler case of bounded open subsets with smooth boundary in $\mathbb{R}^2$. It generalizes steadily to all the other situations in which Schwarz symmetrization can be used. Some of them are described at the end of this section.

For symmetrization, we have used the basic reference [H-L-P] pages 260 to 299.

## The Geometric Symmetrization.

From now on the domain under consideration will be connected, bounded and with smooth boundary in $\mathbb{R}^2$.

The symmetrization is a map which associates to each such domain a more symmetric one. The Schwarz symmetrization (which is the one we consider here) associates to $\Omega$ the ball $\Omega^*$ of $\mathbb{R}^2$ with center 0 and the same area as $\Omega$.

Notice that we consider concentric balls in order to have a totally ordered family of balls as target space for the symmetrization.

The main feature of this operation is given by the

**21. Theorem**

*With the above notations if $L$ (resp. $L^*$) is the length of the boundary of $\Omega$ (resp. $\Omega^*$), then*

$$L \geq L^*.$$

22. Remarks.

i) This is the classical isoperimetric inequality in dimension 2. For a review on the different proofs, see [B-Z] and [ON].

ii) We will see later on that this inequality implies all the inequalities which will appear in this section.

iii) Here the smoothness of the boundary is not essential.

23. Some elementary properties

Let A and B be two bounded measurable sets then:

i) $\mathrm{Vol}([A\cap B]^*) = \mathrm{Vol}(A\cap B) \leq \mathrm{Vol}(A^*\cap B^*) = \mathrm{Min}\{\mathrm{Vol}(A),\mathrm{Vol}(B)\}$.

ii) $\mathrm{Vol}([A\cup B]^*) = \mathrm{Vol}(A\cup B) \geq \mathrm{Vol}(A^*\cup B^*) = \mathrm{Max}\{\mathrm{Vol}(A),\mathrm{Vol}(B)\}$.

iii) The symmetrized sets of a finite sequence of sets $(A_k)$ can always be arranged into a decrasing sequence. So, this map is often called "decreasing rearrangement" (see [H-L-P]).

Symmetrization of Functions.

Let A be a measurable set of finite volume then $\chi_A$, the characteristic function of A, is measurable and integrable. We define,

$$S(\chi_A) = \chi_{A^*}.$$

The finite sums of characteristic functions of measurable sets is dense in the space of integrable functions on a bounded domain. Unfortunately the map S cannot be extended as a linear map indeed

$$S(\chi_A+\chi_B) \neq (S(\chi_A) + S(\chi_B))$$

in general.

However, if $A \subset B$ or $B \subset A$, we can define $S(\chi_A+\chi_B)$ by

(*) $$S(\chi_A+\chi_B) = S(\chi_A) + S(\chi_B).$$

Let $\Omega$ be a bounded domain in $\mathbb{R}^2$ and $f$ be a non-negative integrable function on $\Omega$. We will write $f$ as a sum (integral) of characteristic functions of an increasing family of bounded measurable sets.

Define

$$D_t = \{x \in \Omega : f(x) \geq t\}$$

the set $D_t$ is measurable of finite volume.

24. Lemma:

<u>With the above notations</u> $f = \int \chi_{D_t} \, dt.$

<u>Proof</u>. Define the functions

$$F: \Omega \times \mathbb{R}_+ \to \mathbb{R}_+$$

by

$$F(x,t) = \chi_{D_t}(x)$$

and

$$\phi: \Omega \times \mathbb{R}_+ \to \mathbb{R} \quad \text{by} \quad \phi(x,t) = f(x)-t \;.$$

Then

$$F(x,t) = \chi_E(x,t) \quad \text{where} \quad E = \phi^{-1}(\{y \geq 0\}).$$

The function $\phi$ is measurable and so is $E$, thus $F$ is a measurable and non-negative function.

Then by Fubini's Theorem (see [RN] pg. 140) both functions

$$t \to \int_{\Omega} \chi_{D_t}(x)dx, \quad x \to \int_{\mathbb{R}_+} \chi_{D_t}(x)dt = \int_0^{f(x)} 1\, dt = f(x)$$

are measurable and

$$\int_{\mathbb{R}_+} \mathrm{Vol}(D_t)dt = \int_{\mathbb{R}_+} \Big(\int_{\Omega} \chi_{D_t}(x)dx\Big)dt = \int_{\Omega} \Big(\int_{\mathbb{R}_+} \chi_{D_t}(x)dt\Big)dx = \int f < \infty$$

The family $D_t$ is a decreasing family of bounded measurable sets; taking (*) into account we define

$$S(f) = S\Big(\int \chi_{D_t} dt\Big) = \int S(\chi_{D_t})dt = \int \chi_{D_t^*}\, dt.$$

Now it remains to verify that if $f \in L^1(\Omega)$, i.e. is a class of functions defined up to measure zero sets, so is $S(f)$.

More precisely, if $g$ is in the class of $f$ (defines the same $L^2$-function) then $f$ and $g$ differ on a set $N$ of measure zero. Then if

$$E_t = \{x \mid g(x) \geq t\}$$

$$(E_t \setminus D_t) \cup (D_t \setminus E_t) \subset N$$

and has measure zero. Therefore

$$\mathrm{Vol}(E_t) = \mathrm{Vol}(D_t)$$

and

$$E_t^* = D_t^* \quad \text{for all} \quad t$$

which implies

$$S(f) = S(g). \blacksquare$$

For another approach of the symmetrization of functions the reader is referred to Chapter IV, §A, and Chapter V, nº 12.

If $f$ is any integrable function (not necessarily non-negative), we define:

$$S(f) = S(|f|).$$

The next theorem shows why $S$ deserves its name.

25. <u>Theorem</u>:

<u>With above notations the map</u> $S$ <u>is a symmetrization in the sense of Section</u> III.

<u>Proof</u>. The domain $\Omega$ being bounded, any function $f$ in $L^2(\Omega,\mathbb{R})$ is integrable, so $S$ is defined on $L^2(\Omega,\mathbb{R})$.

The symmetrized function $S(f)$ is radially symmetric on $\Omega^*$, i.e. it depends only on the distance to the origin, and is non-increasing. So, the target Hilbert space is the space $L^2([0,R],rdr)$ (or equivalently $L^2_o(\Omega^*,dx)$, the set of functions in $L^2(\Omega^*,dx)$ which are radially symmetric) where $R$ is the radius of $\Omega^*$.

The cone $\mathcal{K}^+$ is the cone constituted by the non-increasing functions (i.e. functions which are in $L^2$ and whose derivative in the distribution sense is a non-positive measure). It is clearly not self-dual.

Let us then verify the conditions which appear in the definition of a symmetrization.

i) If $f$ and $g$ are non-negative function in $L^2(\Omega;\mathbb{R})$

$$f = \int \chi_{D_t} dt \qquad\qquad g = \int \chi_{E_s} ds$$

then

$$fg = \int \chi_{D_t} \chi_{E_s} dtds = \int \chi_{D_t \cap E_s} dtds$$

so

$$\int_\Omega fg = \int_{\mathbb{R}^2} \mathrm{Vol}(D_t \cap E_s) dtds ;$$

similarly

$$\int_\Omega S(f)S(g) = \int_{\mathbb{R}^2} \mathrm{Vol}(D_t^* \cap E_s^*) dtds$$

and from the property 23.i)

$$\langle f|g\rangle \leq \langle S(f)|S(g)\rangle.$$

26. <u>Remark</u>.

If the reader is not satisfied with the formula proved in Lemma 26 and used in the following calculations, he can use the following property of non-negative measurable functions $f$ on a $\sigma$-finite measured space (see [RN] page 15): there exists an increasing sequence of simple functions $s_n$

$$0 \leq s_1 \leq s_2 \leq \ldots \leq s_n \leq f$$

which converges to $f$ at any point. Recall that a simple function is a finite linear combination of characteristic functions of measurable sets; in the case at hand, the sets are in the family $\{D_t\}_{t \in R}$. Then by Lebesgue's dominated convergence theorem and the uniform continuity of the symmetrization on $L^2(\Omega;\mathbb{R})$ we can avoid using the integral representation.

ii) Now if $f$ is a non-negative real valued function on $\Omega$ and $F$ a non-decreasing function on $\mathbb{R}_+$, we can write

$$F \circ f = \int \chi_{D_t} dF$$

the integral being a Stieljes integral ($F$ is of bounded variation). Then using the positivity of the derivative of $F$ or the method given in Remark 26, one can easily prove that

$$S(F \circ f) = \int S(\chi_{D_t}) dF = \int \chi_{D_t^*} dF = F \circ (S(f)).$$

Then let $f$ be a function in $L^2(\Omega;\mathbb{R})$ and $g$ in $\mathcal{K}^+$, $g$ is in $L^2([0,R],xdx)$ and is non-increasing. For each $t \in \mathbb{R}_+$, the set $D_t^* = (\{x|f(x)| > t\})^*$ is a ball of radius $r(t)$, $0 \leq r(t) \leq R$. The function $r$ is non-increasing. For the sake of simplicity (and this will be sufficient for the sequel) we assume that $r$ is a homeomorphism onto $[0,R]$.

If we define the function $h$ to be $g \circ r$, $h$ is a non-decreasing function and by the previous formula

$$S(h \circ |f|) = h \circ S(|f|) = g.$$

The last equality being achieved because $r$ is assumed to be a homeomorphism. Indeed in that case

$$S(|f|)(r_o) = r^{-1}(r_o)$$

$$h \circ S(|f|)(r_o) = g(r_o).$$

The functions $h \circ |f|$ and $|f|$ having the same level sets,

$$\int_\Omega |f| . h \circ |f| = \int_{\Omega^*} S(f) g.$$

If we now define

$$\operatorname{sign}(f) = \begin{cases} \dfrac{f(x)}{|f(x)|} & \text{if} \quad f(x) \neq 0 \\ 1 & \text{if} \quad f(x) = 0 \end{cases}$$

$$\bar{f} = (\text{sign } f) h_0 |f|$$

we have the equalities,

$$S(\bar{f}) = g$$

$$\langle f | \bar{f} \rangle = \int_\Omega f\bar{f} = \int_{\Omega^*} S(f) g = \langle S(f) | g \rangle$$

which is the pairing condition for the particular case where $r$ is a homeomorphism. ■

27. Remarks.

1) The verification of the pairing condition for a general situation ($r$ not necessarily a homeomorphism) being not necessary we limit ourselves to the above case (this will be clear in the next theorem).

2) The reader can easily verify that this definition of symmetrization coincides with the usual one (see [BE] pp. 47) when the functions under considerations are sufficiently regular.

We can now formulate the main theorem of this section. Let $\Delta$ be the Laplace operator on $L^2(\Omega;\mathbb{R})$ (resp. $\Delta^*$ on $L^2(\Omega^*;R)$) with Dirichlet boundary condition. Recall that it is associated to the quadratic form

$$q(u) = \int_\Omega |du|^2$$

with domain

$$\mathcal{D}(q) = H_o^1 = \{u \in L^2(\Omega;\mathbb{R}) \,/\, \int_\Omega |du|^2 < +\infty,\ u|_{\partial\Omega} \equiv 0\}.$$

Notice that if $u$ is a function in $L^2(\Omega;\mathbb{R})$ which satisfies $\int_\Omega |du|^2 < +\infty$ then $u|_{\partial\Omega}$ is a function in $H^{1/2}(\partial\Omega)$ and so at least in $L^2(\partial\Omega)$ (see [C-P] pg. 101). The equality $u|_{\partial\Omega} \equiv 0$ has to be understood in this sense. We have,

28. <u>Theorem</u>

<u>With the above notations the semigroup</u> $e^{-t\Delta^*}$ <u>dominates the semigroup</u> $e^{-t\Delta}$ <u>for all</u> $t > 0$.

<u>Step 1</u>.

The theorem will be proved if we show that for any $f_1 \in L^2(\Omega)$ and $g \in K^+$ we have the inequality

$$(*) \qquad \langle S[(\Delta+\lambda)^{-1}f_1] | g\rangle \leq \langle (\Delta^*+\lambda)^{-1}S(f_1) | g\rangle \quad \text{for all} \quad \lambda > 0$$

(See Theorem 16).

The operators $(\Delta+\lambda)^{-1}$ and $(\Delta^*+\lambda)^{-1}$ are positivity preserving (this is a well known fact which can be proved by using the Beurling-Deny criterion). Thus we have,

$$|(\Delta+\lambda)^{-1}f_1| \leq (\Delta+\lambda)^{-1}|f_1|$$

(inequality between functions)

and $|f_1|$ being non-negative,

$$0 \leq (\Delta+\lambda)^{-1}|f_1|.$$

Using the facts that

$$S[(\Delta+\lambda)^{-1}f_1] = S[|(\Delta+\lambda)^{-1}f_1|]$$

and

$$S(u) \leq S(v) \quad \text{whenever} \quad 0 \leq u \leq v,$$

we see that it is sufficient to prove the inequality (*) for $f_1$ a non-negative function.

Step 2.

The operators $(\Delta+\lambda)^{-1}$, $(\Delta^*+\lambda)^{-1}$ and $S$ being continuous (for the $L^2$ norms; see proposition 8) it suffices to prove (*) for $g$ in a dense subset of $\mathcal{H}^+$ and $f_1$ in a dense subset in $\mathcal{H}^+$ (the set of non-negative functions of $L^2(\Omega;\mathbb{R})$).

The proof of Theorem 16 relies on the inequality

$$(**)\qquad \langle \Delta h_1 | h_2 \rangle \geq \langle S(h_1) | \Delta^* S(h_2) \rangle$$

for $h_1 = (\Delta+\lambda)^{-1} f_1$, $S(h_2) = (\Delta^*+\lambda)^{-1} g$ and $(h_1, h_2)$ $S(h_2)$-paired. If $f_1$ is a non-negative function then $h_1$ is also non-negative, the operator $(\Delta+\lambda)^{-1}$ being positivity preserving.

It is then clear that it is sufficient to prove (**) for $S(h_2)$ in a core for $\Delta^*$ and $h_1$ in a dense subset of $\mathcal{D}(\Delta) \cap \mathcal{H}^+$ (for the $\mathcal{D}(\Delta)$ topology).

Step 3.

We choose for $S(h_2) = u$ a smooth function in $\bar{\Omega}^*$ vanishing on the boundary, radially symmetric and non-increasing in the radial variable.

Then, we have seen that

$$h_2 = g \circ r \circ h_1 \quad \text{where} \quad r(t) = \text{radius of } D_t^*$$

whenever $r$ is continuous. In fact, we will choose $h_1$ to be smooth non-negative and such that $r$ is piecewise smooth and absolutely continuous. Let us assume for a while that this can be done, and define

$$\bar{g} = g \circ r$$

$$h_2 = \bar{g} \circ h_1$$

and $\bar{g}$ is non-decreasing. Thus

$$\langle \Delta h_1 | h_2 \rangle = \int_\Omega (dh_1 | dh_2) = \int_\Omega (\bar{g}' \circ h_1) |dh_1|^2 .$$

Let us define a non-decreasing function on $[0,s]$ (where $s = \sup h_1$), $k$ by

$$\bar{g}' = (k')^2$$

then

$$\langle \Delta h_1 | h_2 \rangle = \int_\Omega (k' \circ h_1)^2 |dh_1|^2 = \int_\Omega |dw|^2$$

with $w = k \circ h_1$.

Now it is a well known fact that symmetrization decreases the Dirichlet integral (see Chap. V, §A or [BE] p. 55)

$$\int_\Omega |dw|^2 \geq \int_\Omega |dS(w)|^2$$

but $S(w) = k \circ S(h_1)$ and by the same process we get

$$\int_\Omega |dw|^2 \geq \langle S(h_1) | \Delta^* S(h_2) \rangle$$

and the theorem is proved.

Step 4.

1) It remains to choose nice functions for $h_1$. Recall that $f_1 = (\Delta+\lambda)h_1$ must be in a dense subset in $\mathcal{H}^+$. For example we can assume that $f_1$ is a smooth non-negative function with compact support in the interior of $\Omega$. Then $h_1$ is smooth up to the boundary ($h_1 \in C^\infty(\bar{\Omega})$), vanishes on $\partial\Omega$ and is positive in the interior of $\Omega$. Thus if $\frac{\partial}{\partial \nu}$ is the derivative in the direction of the inward normal, then

$$\frac{\partial h_1}{\partial \nu} \geq 0.$$

Now we can perturb $h_1$ in $C^\infty(\bar{\Omega})$ in such a way that the new function, let us say $u$, has the following properties

$$\frac{\partial u}{\partial n} > 0 \quad \text{on} \quad \partial\Omega$$

$$u = 0 \quad \text{on} \quad \partial\Omega$$

$$u > 0 \quad \text{in the interior of } \Omega.$$

2) Let $\Omega_s = \{x \in \Omega \setminus \text{dist}(x,\partial\Omega) \geq s\}$. For sufficiently small $s$ this set has a smooth boundary. Rescaling $u$ if necessary we can assume, for the sake of simplicity, that $\frac{\partial u}{\partial s} \geq 1$ on $\Omega\setminus\Omega_{s_o}$ for sufficiently small $s_o$ where $s(x) = \text{dist}(x,\partial\Omega)$.

Furthermore, the function $u$ being in $C^\infty(\bar{\Omega})$ is the restriction on $\bar{\Omega}$ of a smooth, compactly supported, function in $\mathbb{R}^2$.

3) Applying Milnor's theorem (see [MR], page 37) to this extended function, we can approximate uniformly $u$ by a smooth function with non-degenerate critical points on $\mathbb{R}^2$ and in the $C^k$ topology on $\bar{\Omega}$, (for $k$ chosen arbitrarily and $\geq 3$). Let $v$ be such a function

$v \in C^\infty(\bar{\Omega})$

$v$ has non-degenerate critical points

$\|u-v\|_{C^k(\bar{\Omega})} \leq \eta/10$ (small positive number)

if $\eta$ is small enough

$\frac{\partial v}{\partial s} \geq \frac{1}{2}$ on $\Omega\setminus\Omega_{s_o}$ (and so has no critical points in this set).

From the properties of $u$ it is clear that the set

$$\{x\backslash u(x) = \eta\} = \Gamma_\eta$$

is a smooth curve for $\eta$ small which converges to $\partial\Omega$ as $\eta$ goes to zero and which is almost parallel to $\partial\Omega$.

Taking $\eta$ small, this curve is in $\Omega\backslash\Omega_{s_o}$. Thus

$$u(x) = \eta \quad \text{on} \quad \Gamma_\eta \quad \text{implies} \quad v(x) \geq \frac{9\eta}{10} \quad \text{on} \quad \Gamma_\eta$$

$$u(x) = 0 \quad \text{on} \quad \partial\Omega \quad \text{implies} \quad v(x) \leq \frac{\eta}{10} \quad \text{on} \quad \Gamma_\eta.$$

Then the set $\{x\backslash v(x) = \frac{\eta}{5}\} = \gamma_\eta$ is a smooth curve close to $\partial\Omega$.

It now suffices to construct a diffeomorphism $\psi_\eta$ from the interior of $\gamma_\eta$ onto $\Omega$ such that:

$$\begin{cases} \psi_\eta \text{ is the identity in } \Omega_{s_o} \\ \psi_\eta \text{ sends diffeomorphically } \gamma_\eta \text{ onto } \partial\Omega \\ \psi_\eta \text{ is close to the identity in the } C^k \text{ topology.} \end{cases}$$

Then the function

$$w(x) = v[\psi_\eta^{-1}(x)] - \eta/5 \quad \text{for} \quad x \in \Omega$$

is a smooth function, with finitely many critical points in the interior of $\Omega$, vanishing on $\partial\Omega$ and arbitrary close to $h_1$ in the $\mathcal{D}(\Delta)$-topology.

4) We can then work with such functions, which clearly have the property that the associated function $r(t)$ is absolutely continuous (the only possible points at which $r$ is not smooth are the critical values of $h_1$).

Finally we have to verify that $(\Delta^* + \lambda)^{-1}$ preserves $\mathcal{K}^+$. Recall that if $g \in \mathcal{K}^+$, it is radially symmetric. Because the rotations about the origin which are isometries of $\mathbb{R}^2$, commute with $\Delta^*$, $f = (\Delta^*+\lambda)^{-1}g$ is radially symmetric.

The function $f$ verifies (by definition)

$$(\Delta^*+\lambda)f \geq 0$$

thus by applying the maximum principle we see that it is non-increasing in the radial variable. ■

29. Remarks.

i) The technical details have not been completely written here, they will appear elsewhere. We just wanted to show that although the criterion given in theorem 28 is not so beautiful as Beurling-Deny's one, it can be improved in some particular cases; it is sufficient to prove that the energy integral is not increased by symmetrization.

ii) This proof steadily generalizes to higher dimension.

iii) It also gives an alternative proof of Theorem 9 and 28 of Chapter V. In fact, it is much simpler technically in the case of a compact manifold without boundary. Let $(M,g)$ be a $n$-dimensional compact connected Riemannian manifold without boundary such that

$$\mathrm{Ricci}(g) \geq (n-1)g$$

and define the number $\beta$ by,

$$\beta = \frac{\mathrm{Vol}(M)}{\mathrm{Vol}(S^n)} \leq 1$$

then the geometric symmetrization associates to each measurable set $D$ on $M$ a ball centered at the north pole of $S^n$ of volume $\frac{1}{\beta}\mathrm{Vol}(D)$. Gromov's isoperimetric inequality ([GV]) asserts that this map decreases the volume of $\partial D$ (when it is smooth) up to the factor $\beta$. This yields a symmetrization on functions

$$f = \int \chi_{D_t} dt \qquad S(f) = \int \chi_{D_t^*} dt ;$$

one then has to prove

$$\beta^{-1} \int_M |dw|^2 \geq \int_{S^n} |dS(w)|^2$$

for nice $w$ which is easy because we deduce from Milnor's theorem that the functions with non-degenerate critical points are dense in $C^k(M)$ for all $k \in \mathbb{N}$.

iv) The theorems on forms which appear in [B-G] are obtained by composing symmetrizations.

## REFERENCES FOR APPENDIX A


[AN] T. AUBIN, Non linear analysis on Manifolds. Monge-Ampère équations. Grundlehren der mathematischen Wissenschaften 252. Springer-Verlag.

[BE] C. BANDLE, Isoperimetric inequalities and applications. Monographs and Studies in Mathematics, nº 7. Pitman.

[B-G] P. BERARD et S. GALLOT, Inégalités isopérimétriques pour l'équation de la chaleur et applications à l'estimation de quelques invariants.

[B-Z] Ju. D. BURAGO & V.A. ZALGALLER, Geometric Inequalities, Nauka, Leningrad 1980 (Russ.)

[C-P] J. CHAZARAIN & A. PIRIOU, Introduction à la théorie des équations aux dérivées partielles linéaires collection $\mu_B$. Gauthiers-Villars.

[GV] M. GROMOV, Paul Levy's isoperimetric inequality. Prépublication I.H.E.S. 1980.

[H-L-P] HARDY, LITTLEWOOD & G. POLYA, Inequalities. Cambridge University Press.

[H-S-U] H. HESS, R. SCHRADER & D.A. UHLENBROCK, Kato's inequality and the spectral distribution of Laplacians on compact Riemannian manifolds. J. Diff. Geo. 15 (1980), 27-37.

[H-S-U1] H. HESS. R. SCHRADER & D.A. UHLENBROCK, Domination of semigroups and generalization of Kato's inequality. Duke Math. J. 44 (1977) 893-904.

[KO] T. KATO, Schrödinger operators with singular potential. Israel J. Math. 13 (1973), 135-148.

[MR] J. MILNOR, Morse theory. Annals of Mathematical studies. Study 51. Princeton University Press.

[ON] OSSERMAN, The Isoperimetric Inequality. Bull. A.M.S. 84 (1978), 1182-1238.

[RN] W. RUDIN, Real and complex analysis. McGraw-Hill series in higher Mathematics.

[R-S]IV M. REED & B. SIMON, Methods of modern mathematical physics. Vol. IV. Academic Press.

[SN] B. SIMON, Kato's inequality and the comparison of semigroups. J. of Funct. Analysis 32 (1979) 97-101.

# APPENDIX B

## A GUIDE TO THE LITERATURE

"Le Spectre d'une Variété Riemannienne en 1982"

by

Pierre Bérard and Marcel Berger

This bibliography was first published in "Spectra of Riemannian Manifolds" Kaigai Publications, Tokyo 1983, p. 139-194.

Spectra of Riemannian Manifolds
Kaigai Publications, Tokyo, 1983, 139-194

# LE SPECTRE D'UNE VARIÉTÉ RIEMANNIENNE EN 1982

PIERRE H. BÉRARD & MARCEL BERGER

## Table des Matières



## Introduction

Depuis 1970, date à laquelle a été publié "LE SPECTRE D'UNE VARIÉTÉ RIEMANNIENNE", en abrégé, BGM (M. Berger, P. Gauduchon, et E. Mazet, Lecture Notes in Mathematics, n° 194, Springer) l'étude du spectre a connu une grande effervescence. Il nous a paru utile de rassembler une *bibliographie classée et assez complète* (mais bien sûr difficilement exhaustive) pour compléter le BGM.

Suivant en cela BGM, mais aussi pour des raisons de temps, d'espace et d'incompétence, nous avons fait cette bibliographie avec un A PRIORI: quand nous disons SPECTRE, nous sous-entendons "*le spectre du Laplacien d'une variété riemannienne compacte sans bord*".

Il ne nous a cependant pas paru raisonnable de nous limiter à ce seul sujet; c'est pourquoi nous donnons aussi des éléments de bibliographie concernant le spectre de l'opérateur de Laplace-Beltrami agissant sur les p-formes (essentiellement regroupés au paragraphe 3.2) et la théorie spectrale des variétés non compactes (voir chapitre 10). Compte tenu de l'importance "physique" des variétés à bord, et aussi des développements assez spectaculaires dont la théorie a été l'objet ces dernières années, nous donnons un certain nombre de références concernant le "*cas à bord*". Ces références sont ventilées dans les différents chapitres. Notons que, dans les chapitres 3 à 8, et 10, ces références sont regroupées à la fin de chaque paragraphe, précédées de la mention explicite *"cas à bord"*.

Peut-être convient-il de noter que le cas des variété non compactes, comme celui des variétés à bord, sort du cadre strict de la géométrie riemannienne (problèmes de théorie spectrale dans l'un, de géométrie symplectique dans l'autre).

Le lecteur trouvera dans le *"Mode d'emploi"* ci-après des informations plus détaillées qui l'aideront, du moins nous l'espérons, à utiliser fructueusement cette bibliographie.

Plusieurs collègues ont bien voulu nous signaler des erreurs, des omissions ou de nouvelles références; nous les prions de bien vouloir accepter ici nos remerciements.

Nous remercions Mesdames Cordel et Strazzanti qui ont assuré la frappe du manuscrit.

## Mode d'Emploi

*"Le spectre en 1982"* est divisé en douze chapitres. Les chapitres 1 et 2 sont essentiellement consacrés aux résultats préliminaires. Le chapitre 9 traite des VARIÉTÉS SPÉCIALES et le chapitre 10 des VARIÉTÉS NON COMPACTES. Notons que les références concernant certains sujets *pointus* (et souvent encore peu explorés) ont été regroupées, sous différentes rubriques au chapitre 12. Nous renvoyons le lecteur au *"Leitfaden"* pour plus de détails concernant le contenu des différents chapitres.

Quelques commentaires complémentaires:

*"cas à bord"*: dans les chapitres 3 à 8 et 10, cette mention précède, à la fin de chaque paragraphe, les références concernant le spectre des variétés à bord.

*"Tableau des interactions fortes"*: toutes les références bibliographiques n'ont pas été inscrites dans chacun des paragraphes où elles devraient l'être. Ce tableau est destiné à compenser cet inconvénient.

*"Ouvrages de base"*: nous avons choisi (choix personnel, donc sujet à caution) un certain nombre de références: livres, cours, articles de synthèse, pour aider le lecteur à se faire une idée générale d'un chapitre précis, avant d'aborder la jungle de la bibliographie spécialisée. Pour être plus repérables, ces références sont données en MAJUSCULES (exemples: CLARK [1], GUILLEMIN [2]). Une liste spécifique de ces ouvrages est donnée après le *"Tableau des interactions fortes"* (chaque référence est suivie des numéros des paragraphes auxquels elle se rapporte).

*"Actes de colloques"*: il est parfois intéressant de connaître le développement historique d'un sujet. C'est pourquoi nous avons regroupé, en une liste séparée, les actes des colloques où sont publiés certains des articles cités en référence (ceci par ordre chronologique).

*"Preprints"*: pour permettre au lecteur de localiser (ou de se procurer) plus facilement les articles encore sous forme de prétirage, nous avons essayé de donner un lieu d'émission et une date (en général celle à laquelle nous avons eu le prétirage en main pour la première fois; cette date, on s'en doute, est assez relative et on ne peut lui donner de valeur absolue).

*"Validité"*: cette bibliographie a été arrêtée à l'état de nos fiches en septembre 81. La composition du texte ayant été retardée, cela nous a amenés à compléter notre texte un an plus tard, soit en décembre 82. Nous osons espérer que le lecteur nous pardonnera de ne pas avoir respecté l'ordre lexicographique dans la classification par sujets.

Dans le "Leitfaden" qui suit, nous donnons quelques indications complémentaires sur la manière dont nous avons ventilé les références suivant les chapitres.

Terminons en indiquant trois références qui peuvent être utilisées par le lecteur parallèlement à cette bibliographie:

Simon-Wissner [1]: article de synthèse sur une partie des chapitres qui constituent cette bibliographie;

Yau [3]: le lecteur y trouvera une liste des applications des équations aux dérivées partielles à la géométrie et en particulier au spectre;

Yau [4]: liste de problèmes ouverts en géométrie, contient une section propre au spectre (certains problèmes sont peut-être déjà résolus).

## Leitfaden

Chapitre 1: *"Préliminaires à l'étude du spectre"*

Ce chapitre est surtout destiné aux non spécialistes. Nous y donnons quelques références (personnelles) sur les connaissances requises pour aborder la littérature spécialisée.

Chapitre 2: *"Motivations; Equations de la physique mathématique"*

L'intérêt porté au spectre du Laplacien nous vient sans doute de la physique. Les références concernant les rapports avec la physique sont données au paragraphe 2.1. Dans le paragraphe 2.2. nous donnons des références relatives à l'étude a priori des équations de la physique mathématique utiles dans l'étude du spectre.

Chapitre 3: *"Exemples de spectres"*

Comme le montrent les références données dans le paragraphe 3.1., les variétés dont le spectre est donné par des formules explicites sont rares.

On a cependant une bonne description du spectre de certaines variétés spéciales (groupes de Lie, espaces symétriques, ...), parfois par le biais de solutions explicites pour l'équation de la chaleur ou des ondes. Nous donnons ces références dans le paragraphe 3.3.

Nous avons choisi de rassembler toutes les références relatives au spectre de l'opérateur de Laplace-Beltrami agissant sur les formes différentielles dans le paragraphe 3.2. (en particulier, elles ne sont pas ventilées systématiquement dans les différents paragraphes). C'est une conséquence de l'A PRIORI cité dans l'introduction.

Dans ce chapitre, les références au *"cas à bord"* sont systématiquement regroupées en fin de chaque paragraphe.

Chapitre 4: *"Asymptotiques"* et Chapitre 5: *"Spectre et géométrie"*

De très nombreux résultats sur le spectre ont été obtenus par l'intermédiaire du comportement asymptotique de certaines fonctions du spectre.

L'équation fonctionnelle de l'exponentielle a permis, par le biais de l'asymptotique du noyau de la chaleur (à la Minakshisundaram-Pleijel) d'étudier certains invariants spectraux: paragraphes 4.1. et 5.1. respectivement.

L'étude de la propagation des ondes, liée aux géodésiques (en fait aux trajectoires d'un hamiltonien) a permis d'établir les rapports existant entre le spectre et le spectre des longueurs, que sont les formules de Poisson: paragraphes 4.2. et 5.2. (à comparer aussi avec les formules de traces de Selberg: paragraphe 9.1.).

Ces études ont permis de mieux cerner le comportement asymptotique des valeurs propres: paragraphe 4.3.

Les paragraphes 5.4. et 5.5. sont consacrés aux références relatives à des sujets connexes.

La fonction zeta associée aux valeurs propres joue aussi un rôle important, le paragraphe 5.3. lui est consacré.

Dans ces deux chapitres, les références au *"cas à bord"* sont regroupées à la fin de chaque paragraphe.

Chapitre 6: *"Isospectralité"*

Ce sujet, presque intouché en 1970, a connu d'importants développements récents. Le paragraphe 6.1. est consacré aux résultats positifs (souvent très liés au paragraphe 5.1.): zoologie des variétés caractérisées par leur spectre, et à certains théorèmes généraux.

Le paragraphe 6.2. est consacré à la faune des variétés isospectrales non isométriques.

Ici encore, le *"cas à bord"* fait l'objet d'attentions particulières.

Chapitre 7: *"Perturbations et généricité"*

Outre les références relatives aux propriétés génériques du spectre (paragraphe 7.2.) ce chapitre contient des références sur le comportement du spectre sous divers types de perturbations (paragraphe 7.1.).

Les références pour le *"cas à bord"* sont données à la fin de chaque paragraphe.

Chapitre 8: *"Equations aux dérivées partielles, applications"*

Le fait que le spectre étudié soit celui d'un opérateur différentiel impose des conditions locales ou globales sur les fonctions propres et par conséquent sur le spectre lui-même. De même le fait que le Laplacien soit très lié à la structure riemannienne conduit à des utilisations spécifiques des fonctions propres (immersions isométriques ...). Des références à ces divers aspects de l'étude du spectre sont données dans ce chapitre.

Les références pour le *"cas à bord"* sont données à la fin de chaque paragraphe.

Chapitre 9: *"Variétés spéciales"*

Le paragraphe 9.1. est consacré au cas très particulier des variétés de courbure $-1$: techniques et résultats sont propres à la géométrie hyperbolique (comme par exemple la formule des traces de Selberg) mais la comparaison avec le "cas général" n'en est pas moins intéressante.

Le paragraphe 9.2. contient les références qui traitent du spectre d'autres variétés particulières.

Chapitre 10: *"Cas non compact"*

Selon l'A PRIORI indiqué dans l'introduction, nous avons regroupé les références relatives à l'étude du spectre des variétés non compactes en un seul chapitre, sans les ventiler en différents paragraphes. Le *"cas à bord"* fait quand même l'objet d'un traitement séparé.

Chapitre 11: *"Etude individuelle des valeurs propres"*

La première valeur propre (non triviale) joue un rôle particulier (comme en physique), le paragraphe 11.1. lui est consacré.

Les autres valeurs propres se contentent du seul paragraphe 11.2.

Le paragraphe 11.3. est consacré aux questions connexes: inégalités isopérimétriques et inégalités de Sobolev. Ces questions sont liées à l'étude des valeurs propres. Ce texte *n'étant pas* une bibliographie spécifique sur ce sujet, et compte tenu des excellentes références Payne [1], Osserman [2, 3] et Bandle [1, 3] nous ne donnons que certaines références antérieures à ces trois articles et bien sûr celles d'articles plus récents (sans doute en avons nous oubliées!).

Chapitre 12: *"Last but not least"*

Dans ce chapitre nous avons regroupé, sous diverses rubriques, les références des articles qui traitent d'aspects particuliers du spectre. Ce sont souvent des domaines peu explorés, au moins actuellement, ou des domaines connexes au spectre. Aussi, il convient de ne pas considérer ce chapitre comme mineur.

## Le Spectre en 1982

1. PRÉLIMINAIRES À L'ÉTUDE DU SPECTRE
   (Théorie spectrale abstraite; théorie spectrale des opérateurs différentiels; équations aux dérivées partielles; matériel riemannien)
2. MOTIVATIONS: ÉQUATIONS DE LA PHYSIQUE MATHÉMATIQUE
   - 2.1. Motivations; Physique et modèles mathématiques;
   - 2.2. Résultats généraux sur les équations étudiées: problèmes de Dirichlet et de Neumann pour le Laplacien; équation de la chaleur, équation des ondes; fonctions de Green;
3. EXEMPLES DE SPECTRES
   - 3.1. Exemples numériques explicites de spectres ou de valeurs propres du Laplacien sur les fonctions;
   - 3.2. Le Laplacien sur les formes;
   - 3.3. Spectre du Laplacien sur les fonctions et variétés spéciales (groupes de Lie, espaces symétriques, quotients, submersions, ...); expressions "explicites" pour les noyaux de la chaleur et des ondes;
4. ASYMPTOTIQUES
   - 4.1. Développements asymptotiques à la MINAKSHISUNDARAM-PLEIJEL;
   - 4.2. Formules de POISSON et équation des ondes;
   - 4.3. Asymptotique des valeurs propres;
5. SPECTRE ET GÉOMÉTRIE
   - 5.1. Spectres et invariants locaux et globaux;
   - 5.2. Spectre des longueurs; spectre et longueurs des géodésiques périodiques;
   - 5.3. Fonctions zeta; invariant êta;
   - 5.4. Quasimodes; fonctions propres concentrées près d'une géodésique périodique;
   - 5.5. Spectre du Laplacien plus potentiel;
6. ISOSPECTRALITÉ
   - 6.1. Résultats positifs et théorèmes généraux;
   - 6.2. Contre-exemples;

7. PERTURBATIONS ET GÉNÉRICITÉ
    7.1. Perturbations du spectre, du Laplacien, formules de variation à la Hadamard;
    7.2. Résultats sur la généricité;
8. ÉQUATIONS AUX DÉRIVÉES PARTIELLES: APPLICATIONS
    8.1. Etude locale et applications;
    8.2. Etude globale et applications;
9. VARIÉTÉS SPÉCIALES
    9.1. Cas hyperbolique: courbure $-1$; formules de traces de Selberg;
    9.2. Autres variétés spéciales: submersions riemanniennes; espaces lenticulaires; variétés et tores plats; espaces riemanniens symétriques de rang 1; groupes de Lie et quotients discrets; autres;
10. LE CAS NON COMPACT
11. ÉTUDE INDIVIDUELLE DES VALEURS PROPRES
    11.1. Estimées sur le $\lambda_1$, et applications;
    11.2. Estimées faisant intervenir les $\lambda_k$, $k \geqq 2$ et applications;
    11.3. Inégalités isopérimétriques; inégalités de Sobolev; applications;
12. LAST BUT NOT LEAST

Lignes et surfaces nodales; spectre et actions de groupes; approximations et triangulations; calculs numériques approchés; variétés avec singularités; convergence des séries de fonctions propres; opérateurs autres que le Laplacien; invariant êta; torsion analytique; inégalités de type isopérimétrique autres que celles du § 11.3.; géométrie intégrale et problèmes spectraux; multiplicités des valeurs propres; modifications par attachement d'anses; probabilités et géométrie; divers.

## 1. PRÉLIMINAIRES A L'ETUDE DU SPECTRE

Généralités

Cime 1973:3, CLARK [1], FRIEDLAND [2], GARABEDIAN [1], Gelfand [1], Gelfand-Yaglom [1], GOULAOUIC [1]

Théorie spectrale abstraite

Glazman [1], REED-SIMON [1] (vol II)

Théorie spectrale des opérateurs différentiels

BROWDER [1], Protter [2]

Equations aux dérivées partielles

Gilbarg-Trudinger [1], Petrovsky [1], PROTTER [3]

Matériel Riemannien

BERGER-GAUDUCHON-MAZET [1], Besse [1]

2. MOTIVATIONS; ÉQUATIONS DE LA PHYSIQUE MATHÉMATIQUE

Généralités

CLARK [1], Courant-Hilbert [1] (vol 1), Gelfand-Yaglom [1], Kac [1, 2], Morse-Feshbach [1], VENKOV [1]

2.1. Mativations; Physique et modèles mathématiques

Balian-Bloch [1], Petrovsky [1], PROTTER [3]

2.2. Résultats généraux sur les équations étudiées: problèmes de Dirichlet et de Neumann pour le Laplacien; équation de la chaleur, équation des ondes; fonctions de Green

Atiyah-Bott-Patodi [1], Aubin [1], Benabdallah [1], Cheeger-Yau [1], Cheng-Li-Yau [1, 2], Colin de Verdière [3, 7, 8], Colin de Verdière-Frisch [1], Dodziuk [2, 3], Fegan [1, 3], Frisch [1], Greiner [1, 2], GUILLEMIN-STERNBERG [1], Hall-Stedry [1], Hess-Schrader-Uhlenbock [1], Hörmander [2], Ivrii [1, 2], Kannai [1], Keller [1],Keller-Rubinow [1], Lascar [1], Malliavin [1], Meyer [1], Minakshisundaram-Pleijel [1], Mneimne [1], Molchanov [1], Rauch [1], Reilly [5], Seeley [1, 4, 5, 6], Smale [1], Urakawa [4], Weinstein [5], Zucker [1], Arnal [1], Cheng-Li [1], Clements [1], Danet [1], Eichhorn [5], Günther [5], Har'el [2], Kalnins-Miller [1], Oersted [1], Rinke-Wunsch [1], Varopoulos [1 à 4]

3. EXEMPLES DE SPECTRES

Généralités

BERGER [1], Courant-Hilbert [1], Morse-Feshbach [1], Paquet [1]

3.1. Exemples numériques explicites de spectres ou de valeurs propres du Laplacien sur les fonctions

Buser [8], Friedland-Hayman [1], Urakawa [9]

"*cas à bord*": Bérard [4, 6], Bérard-Besson [2], Nooney [1], Pinsky [3], Polya [1], Urakawa [6]

3.2. Le Laplacien sur les formes

Asada [1, 2], Donnelly [19], Eichhorn [3, 7], Fegan [2, 3], Ikeda-Taniguchi [1], Iwasaki-Katase [1], Kuwabara [3], Levy-Bruhl [2, 3], Millman [1], Tachibana-Yamaguchi [1], Tanno [2, 8], Tsagas [1], Tsagas-Kochinos [1], Wolpert [5], Dodziuk [5, 6, 7, 8, 9, 10], Tsukamoto [1]

3.3. Spectre du Laplacien sur les fonctions et variétés spéciales (groupes de Lie, espaces symétriques, quotients, submersions, . . .); expressions "explicites" pour les noyaux de la chaleur et des ondes

Bedford-Suwa [1], Beers-Millman [1], Benabdallah [1], Bérard Bergery- Bourguignon [1, 2], Besson [1], Cheeger-Taylor [1], Chen-Vanhecke [1], S. S. Chen [1], Gallot-Meyer [1], Huber [3, 4, 5], Ikeda [1, 3], Ikeda-Yamamoto [1], Y. Mutō [1, 2], Sakai [2], Strese [1, 4], Sunada [1], Tandai-Sumitomo [1], Taniguchi [1], Tanno [4], Tsagas [2], Yamaguchi [1], Berezin [1], Furutani [1], |Marbes [1, 2], Tsukada [3], Urakawa [9]

4. ASYMPTOTIQUES

Généralités

Balian-Bloch [1, 2, 3], BÉRARD [3], BERGER [1], CLARK [1], Colin de Verdière [3, 8], Duistermaat-Guillemin [1], Duistermaat-Kolk-Varadarajan [1], Gangolli [1], Guillemin [1, 3, 4], GUILLEMIN [2], GUILLEMIN-STERNBERG [1]

*"cas à bord"*: Balian-Bloch [1, 2, 3], CLARK [1], Seeley [4, 5, 6]

4.1. Développements asymptotiques à la MINAKSHISUNDARAM-PLEIJEL

Bérard [1, 2], Cahn-Gilkey-Wolf [1], Chavel-Feldman [4], Cheeger [1], COLIN DE VERDIERE [1, 6], Dlubek-Friedrich [1], Dodziuk [3], Fegan [1], Greiner [1, 2], Hess-Schrader-Uhlenbock [1], Kannai [1], Miatello [1], Minakshisundaram-Pleijel [1], Mneimne [1], L. Smith [1], Wallach [1], Atiyah [1], Bott [1], Sunada [5]

*"cas à bord"*: Hasegawa [1], L. Smith [1], Atiyah [1], Bott [1]

4.2. Formules de POISSON et équation des ondes

Bérard [2], Besse [1], Chazarain [1, 3, 4, 5], CHAZARAIN [2], Colin de Verdière [5, 7], Kolk [1]

*"gas à bord"*: Bardos-Guillot-Ralston [1, 2], Harthong [1], Kurylev [1]

4.3. Asymptotique des valeurs propres

Bérard [5], Boutet de Monvel [1, 4], Boutet de Monvel-Grisvard [1], Boutet de Monvel-Guillemin [1], Chachère [1], Clerc [1, 2], Colin de Verdière [4], Fleckinger-Pellé [1], Frisch [2, 3], GOULAOUIC [1], Grubb [1, 2], Haitov [1], HEJHAL [1], Hejhal [2], Helffer-Robert [1], Helton [1], Hörmander [1, 2, 3], Kolk [1, 2], Lieb [3], Meyer [1], Randol [2, 3], Taylor [1], Vasil'ev [1], VENKOV [1], Weinstein [1, 2, 3, 4], Weyl [1], Widom [1, 2, 3], Asurov [1]

*"cas à bord"*: Arnold J. M. [1], Babich [1], Babich-Levitan [1], Bérard [4, 6], Bérard-Besson [4], Brüning [1], Ivrii

[1, 2, 3, 4, 5], Keller-Rubinov [1], Kurylev [1], Majda-Ralston [1, 2], Melrose [1, 2], Pham The Lai [1], Pinsky [3], Polya [1], Seeley [2, 3], Boimatov-Kostjucenko [1], Carleman [1], Lazutkin-Terman [1], Tamura [1 à 4]

5. SPECTRE ET GÉOMÉTRIE

Généralités

BÉRARD [3], BERGER [1, 3], Fischer [1], GUILLEMIN-STERNBERG [1], Kac [1], Singer [1]

*"cas à bord"*: Kac [1], Fischer [1]

5.1. Spectres et invariants locaux et globaux

Atiyah-Bott-Patodi [1], Benko et al [1], Bérard [1], Brooks [1, 2, 3], Brüning [3], Brüning-Heintze [1], Cahn-Gilkey-Wolf [1], Cheeger [1], COLIN DE VERDIÉRE [6], Dodziuk [3], Dodziuk-Patodi [1], Donnelly [1, 2, 3, 4, 6, 12, 13, 14, 15], Donnelly-Patodi [1], Gallot-Meyer [1], Gilkey [1, 2, 4, 6à23], GILKEY [3], Gilkey-Sacks [1], Greiner [1, 2], Günther-Schimming [1], Har'el [1], Hasegawa [2], Ii [1], Levy-Bruhl [1], Mc Kean-Singer [2], Müller [1], Patodi [1, 2, 3], Olszak [1], Perrone [1], Pinsky [4], Ray [1], Ray-Singer [1, 2], Sakai [1], Sunada [2], Tanno [1], Urakawa [5], Véron [1],

*"cas à bord"*: Gilkey [5], Hasegawa [1], Mc Kean-Singer [1], L. Smith [1], Kennedy [1], Schimming [1], Schimming-Teumer [1]

5.2. Spectre des longueurs; Spectre et longueurs des géodésiques périodiques

Balian-Bloch [1 à 4], Bérard [2], Bérard Bergery [1], Besse [1], Boutet De Monvel [1], Boutet de Monvel-Guillemin [1], Buser [10, 11], Chachère [1], CHAZARAIN [2], Chazarain [1, 3], Colin de Verdière [1, 3, 7, 8], De George [1], Donnelly [9], Duistermaat-Guillemin [1], Frisch [1], Gangolli [2], Guillemin [1, 3, 4], GUILLEMIN [2], Guillemin-Weinstein [1], Helton [1], Kudla-Millson [1], Müller [2], Randol [1, 3, 4, 6], Weinstein [1, 2, 5], Wolpert [3],

*"cas à bord"*: Balian-Bloch [1 à 4], Guillemin-Melrose [1, 2], Harthong [1], Marvizi-Melrose [1], Millson [1]

5.3. Fonctions zeta; invariant êta

Atiyah-Bott-Patodi [1], Atiyah-Patodi-Singer [1], Cahn [1], Cahn-Wolf [1], Dlubek-Friedrich [1], Donnelly [5, 7, 10, 12], Gangolli [3], Gilkey [11], Randol [5], Seeley [1], VENKOV [1], S. Tanaka [1], Wodzicki [1], Atiyah-Donnelly-Singer [1], Gilkey-Smith [1], Millson [1]

5.4. Quasimodes; fonctions propres concentrées près d'une géodé-

sique périodique
Arnold [1], Colin de Verdière [2], Guillemin-Weinstein [1], Pyshkina [1], Ralston [1, 2], Weinstein [5],
*"cas à bord"*: Babich [1], Babich-Lazutkin [1], Keller-Rubinow [1], Lazutkin [1, 2, 3], Babich-Ulin [1] Lazutkin-Terman [2]

5.5. Spectre du Laplacien plus potentiel
Barthel-Kümritz [1], Colin de Verdière [3, 4, 5, 8, 9], Flaschka [1], Fleckinger-Pellé [1], Guillopé [1, 2], Lax-Phillips [1, 2], Majda-Ralston [1], Prosser [2], Weinstein [3, 4], Widom [1, 2, 3], Fegan [4], Fleckinger [1], Moser [3]
*"cas à bord"*: Balian-Bloch [3], Chung-Li [1], Guillemin [5, 6], Guillemin-Uribe [1], Li-Yau [3], Voros [1, 2]

6. ISOSPECTRALITÉ
Généralités—A titre d'exemple (mais un peu en dehors du sujet) FRIEDLAND [4, 5], KITAOKA [1], Calogero [1, 2], Carison [1], Carroll [1], Carroll-Gilbert [1], Carroll-Santosa [1] Chudnovsky-Chudnovsky [1], Levitan [1]

6.1. Résultats positifs et théorèmes généraux
Bérard [1], Berry [1], Buser [10, 11], Donnelly [1, 2, 3], Fischer [1], Flaschka [1], Global Analysis [1], Guillemin-Kazhdan [1, 2], Hochstadt [1], Gilkey [4, 7, 9], Krein [1], Kuwabara [1, 2, 4], Mc KEAN [3] p. 122, Mc Kean-Van Moerbeke [1], Moser [1, 2], Prosser [1, 2], Randol [6], Sakai [1], Sunada [1], Symes [1], M. Tanaka [1, 2], Tanno [1, 6, 7], Wolpert [1, 2, 3], Zalcman [1], Guillemin [5, 6], Marvizi-Melrose [1, 2]
*"cas à bord"*: Borg [1], Guillemin-Melrose [1, 2], Kac [1], Levinson [1], Mallows-Clark [1], Waechter [1], Gelgand-Levitan [1]

6.2. Contre-exemples
Berger-Gauduchon-Mazet [1], Ejiri [1], Ikeda [2, 4, 5], Vignéras [1, 2]
*"cas à bord"*: Hersch [4], Urakawa [8]

7. PERTURBATIONS ET GÉNÉRICITÉ
Généralités
Albert [1 à 4]; Bando-Urakawa [1], Bleecker-Wilson [1], Krein [1], Uhlenbeck [1, 2], Urakawa [6, 7]
*"cas à bord"*: Driscoll [1]

7.1. Perturbations du spectre, du Laplacien, formules de variation à la Hadamard
Aomoto [1], Donnelly [16], Fujiwara [1, 2], GARABEDIAN

[1], Lobo Hidalgo-Sanchez Palencia [1], Rauch [2], Rauch-Taylor [1], Svendsen [1], Tanikawa [2], Wolpert [4], Weber [1]

*"cas à bord"*: Chavel-Feldman [3, 4], Fujiwara et al. [1], Ozawa [1 à 15], Swanson [1], Fujiwara [3], I'lin [1], Maz'ja et al. [1 à 3], Shimakura [1, 2, 3], Vanninathan [1]

7.2. Résultats sur la généricité

Arnold [1], Millman [1]

*"cas à bord"*: Tanikawa [1]

8. ÉQUATIONS AUX DÉRIVÉES PARTIELLES: APPLICATIONS

Généralités

Besson [1], Borell [1], Cheng [1, 2], Cheng-Yau [1], Gallot [1], Huber [6], Tanno [3], Uchiyama [1], Yau [2]

*"cas à bord"*: Brascamp-Lieb [1, 2, 3], Hersch [1]

8.1. Etude locale et applications

Albert [1, à 4], Goldberg-Ishihara [1], H. Muto [2]

*"cas à bord"*: Nooney [1]

8.2. Etude globale et applications

Aubin [2], Brüning [2], Gallot [6], Kobayashi [1], Kobayashi-Takeuchi [1], Li [5], Müller Pfeiffer-Stande [1], Nagano [1], Payne [2], Serrin [1], Takahashi [1]

*"cas à bord"*: Bérard-Meyer [1, 2], Biollay [1], Brüning-Gromes [1], Meyer [1], Peetre [1], Pleijel [1]

9. VARIÉTÉS SPÉCIALES

Généralités

Donnelly [13, 16], Duistermaat-Kolk-Varadarajan [1], Fegan [3], Randol [8], Wolpert [5]

9.1. Cas Hyperbolique: courbure-1, formules de traces de Selberg

Bérard Bergery [1], Buser [1 à 11, 14], BUSER [9], Buzzanca [1], Donnelly [19], Good [1], GUILLEMIN [2], HEJHAL [1], Hejhal [2, 3], Huber [1 à 6], Jenni [1, 2], Kolk [2], Kudla-Millson [1], Lax-Phillips [1, 2, 3], Mc Kean [2], Müller [2, 3], Patterson [1], Randol [1 à 6], Sunada [3], S. Tanaka [1], VENKOV [1], Vignéras [1, 2, 3], Wolpert [1, 3], Ehrenpreis [1], Elstrodt [1, 2], Günther [1, 5], Zograf [1]

9.2. Autres variétés spéciales:

Submersions Riemanniennes

Bérard Bergery [1], Bérard Bergery-Bourguignon [1, 2], Goldberg-Ishihara [1], Y. Mutō [, 2]

Espaces lenticulaires

Ikeda [2], Ikeda-Yamamoto [1], Sakai [2], Tanaka [1, 2]

Variétés et tores plats

Berry [1], Kuwabara [2], Sunada [1], Tsukada [2], Wolpert [2]

Espaces Riemanniens symétriques de rang 1

Bonami-Clerc [1], Bourguignon [1], Cahn-Wolf [1], Gangolli [3], Guillemin [1, 4], Hasegawa [2], Ikeda-Taniguchi [1], Iwasaki-Katase [1], Levy Bruhl [2, 3], H. Muto [1, 2], R. T. Smith [1], Tandai-Sumitomo [1], Tanno [5, 6, 7], Widom [2], Günther [2, 3]

Groupes de Lie et quotients discrets

Beers-Millman [1], Cahn [1], Cahn-Gilkey-Wolf [1], S. S. Chen [1], Clerc [1, 2], Donnely [13], Fegan [1, 2], Urakawa [2, 4], Wallach [1], Berezin [1], Greiner [3], Jerison [1, 2], Nachman [1], Rothschild-Wolf [1]

Autres variétés spéciales

Bedford-Suwa [1], Benabdallah [1], Bérard [1], Chachere [1], Chen-Vanhecke [1], Clerc [3], Colin de Verdière [4], De George [1], Donnelly [1, 2, 3, 17, 21], Donnelly-Li [2], Duistermaat-Kolk-Varadarajan [1], Eichhorn [1], Frisch [1], Gangolli [1, 2], Gilkey [4, 8, 9, 15], Gilkey-Sachs [1], Helgason [1, 2, 3], Ikeda [1, 3], Kashiwara et al. [1], Kuwabara [3], Li [3, 4], Miatello [1], Mneimne [1], Müller [2], Muto-Urakawa [1], Olszak [1, 2], Oshima-Sekiguchi [1], Patodi [3], Sekiguchi [1], Simon [1, 3], Strese [1 à 4], Tandai-Sumitomo [1], Taniguchi [1], Tsagas [2 à 4], Tsukada [1, 4], Urakawa [3, 5, 7], Widom [3], Wolpert [4], Yamaguchi [1], Yang-Yau [1], Bleecker [1], Hano [1], Toimer [1], Yamaguchi [2]

10. LE CAS NON COMPACT

Baider [1], Buser [7], Cheng-Li-Yau [2], Colin de Verdière [5, 9, 10], Donnelly [11, 16, 17, 18, 20, 21], Donnelly-Li [1], Eichhorn [1, 2, 3, 6, 8, 9], Good [1], Guillopé [1], Helffer-Robert [1, 2, 3, 4], Hörmander [3], Jørgensen [3], Mc Kean [1, 2], Müller [2], Randol [8], Sekiguchi [1], Xavier [1], Brooks [1 à 5], Melrose [3]

Asakura [1], Bardos-Guillot-Ralston [1], Jørgensen [1, 2], Majda-Ralston [2], Cantor-Brill [1], Cheeger-Gromov-Taylor [1], Combes-Ghez [1], Dodziuk [4, 6, 7, 8, 9], Elstrodt [1, 2], Elstrodt-Roelke [1], Friedlander [1], Gasimov-Levitan [1], Gehtman [1], Strichartz [1], Vol'pert [1], Voros [1, 2]

11. ÉTUDE INDIVIDUELLE DES VALEURS PROPRES

Généralités

BANDLE [1, 2, 3], Berger [2], Biollay [1], Chavel-Feldman [3], Cheng [1 à 4], Donnelly-Li [3], FRIEDLAND [2], GALLOT [4], Gallot-Meyer [1], GARABEDIAN [1] Garabedian-Schiffer [1], Gromov [1], Hersch [2, 3], Li [1 à 4], Mc Kean [1], OSSERMAN [2], Osserman [3, 4], PAYNE [1], POLYAS-ZEGÖ [1], PROTTER [3], Reid [1], Schoen-Wolpert-Yau [1], Simon [1, 2], Yang-Yau [1], Yau [1], Friedland-Novosad [1], Reilly [6]

11.1. Estimées sur le $\lambda_1$ et applications

Aomoto [1], Asada [1, 2], Aubin [1, 3], Barbosa-do Carmo [1, 2], Barthel-Kümritz [1], Bérard-Besson [1], Bérard-Meyer [1, 2], Bérard Bergery-Bourguignon [1], Berger [4], Besson [2], Bleecker-Weiner [1], Borell [1], Bourguignon [1], Brascamp-Lieb [1, 2, 3], Buser [2 à 8, 12, 14], BUSER [9], do Carmo [1], Chavel [1], Chavel-Feldman [1], Cheeger [3, 5], Chen [1], Croke [1, 2, 5], Debiard-Gaveau-Mazet [1], Friedland [1, 3], Friedland-Hayman [1], Friedrich [1, 2], Fujiwara [1], Gage [1], Gallot [3, 8], Gallot-Meyer [1], Hersch [4], Hoffman [2], Huber [1, 2], Komorowski [1, 2], Li-Treibergs [1], Li-Yau [2], Li-Zhong [1], Marcellini [1], Matsuzawa-Tanno [1], Mazet [1], H. Muto [1, 2, 3], H. Muto-Urakawa [1], Y. Mutō [3], Nehari [1], Obata [1], Osserman [1], Ozawa [6, 8], de Paris [1], Payne-Rayner [1], Philippin [1, 2], Pinsky [4], Protter [1], Randol [1], Reilly [2, 3, 4], Schoen-Wolpert-Yau [2], Sperb [1], Sperner [1], Tachibana-Yamaguchi [1], Tanno [5], M. Taylor [2], Trudinger [1], Tsukada [2], Uchiyama [1], Urakawa [1, 2, 3], Kasue [3], Lichnerowicz [1], Meyer [1], Schoen [1], Sulanke [1], Vignéras [3], Watanabe [1]

11.2. Estimées faisant intervenir les $\lambda_k$, $k \geqq 2$ et applications

Berger [2, 8], Cheng-Li-Yau [1], Gromov [1], Hile-Protter [1], Huber [4], Li-Yau [1], Polya [1], Simon [4, 5], Bareket [1], Cheng-Li [1], Hersch [5], Donnelly-Li [2, 3], Li-Yau [3], Urakawa [10]

11.3 Inégalités isopérimétriques, inégalités de Sobolev, applications

Aubin [2, 4], Barbosa-do Carmo [3], Benko et al. [1], Berger [7, 8], Berger-Kazdan [1], Buser [4, 12], BUSER [9], Chavel-Feldman [2, 5, 6], Cheeger [3], Croke [1], Gallot [2, 5, 7, 9], Gromov [1], Hoffman [1], Ilias [1, 2], Kohler-Jobin [1 à 5], Li-Yau [1], Lieb [1, 2], Peetre [1], Schmidt [1], Cheeger-Gromov-Taylor [1], Chiti [1], Hersch-Monkiewicz [1], I'lin-Moiseev [1], Li [6], Pansu [1, 2], Talenti [1]

## 12. LAST BUT NOT LEAST

Lignes et surfaces nodales

Brüning [2, 4], Brüning-Gromes [1], Cheng [1, 4], Meyer [1], Payne [2], Pleijel [1], Bérard-Meyer [1],

Spectre et actions de groupes

Brüning [3], Brüning-Heintze [1], Donnelly [6, 8, 10, 14], Donnelly-Patodi [1], Gilkey [15, 16], Helgason [1, 2, 3], Höppner [1], Huber [5], R. T. Smith [1], Yen [1], Shafii-Dehabad [1]

Approximations et Triangulations

Dodziuk [1], Dodziuk-Patodi [1], Komorowski [3], Patodi [4]

Calculs numériques

Bassotti Rizza [1], Chachère [1], Forsythe [1]

Variétés avec singularités

Cheeger [4], Cheeger-Taylor [1], Kalka-Menikoff [1]

Convergence des séries de fonctions propres

Alimov et al. [1], Bérard [5], Bonami-Clerc [1], Boutet de Monvel [2], Clerc [1, 2], Hörmander [1], Kenig-Thomas [1], Randol [7], Smale [1], Taylor [1], Meaney [1], Rothschild-Wolf [1]

Opérateurs autres que le Laplacien (opérateurs elliptiques généraux dont opérateur de Dirac; opérateur de Schrödinger...)

J. M. Arnold [1], Balian-Bloch [3], Berthier [1], Boutet de Monvel [3], Boutet de Monvel-Grisvard [1], Boutet de Monvel-Guillemin [1], Buzzanca [1], Chazarain [2, 3], Colin de Verdière [3, 5, 8], Dlubek-Friedrich [1], Flaschka [1], Fleckinger Pellé [1], Friedrich [1], Geller [1], Gilkey [14], Grubb [1, 2], Guillopé [1], Hall-Streedry [1], Kolk [1], Lieb [1, 2, 3], Mc Kean-Van Moerbeke [1], Pyshkina [1], Seeley [1], Strese [2, 3], Sunada [2], Vasil'ev [1], Baxley [1], Bezjaev [1], Charbonnel [1], Cheng [1], Chico [1], Friedrich [2], Kalf [1], Sulanke [1],

Invariant êta

Atiyah-Patodi-Singer [1], Donnelly [5, 7, 10, 12], Gilkey [11], Gilkey-Smith [2], Atiyah-Donnelly-Singer [1]

Torsion analytique

Cheeger [1], Donnelly [12], Müller [1], Ray [1], Ray-Singer [1, 2], Schwarz [1], Urakawa [5]

Inégalités type isopérimétriques autres que §11.3

Berger [5, 6, 7], Berger-Kazdan [1], Kasue [2, 4, 5], Kohler-Jobin [6, 7], Lions [1], Pach [1], Parks [1]

Géométrie intégrale et problèmes spectraux

## Tableau des Interactions Fortes

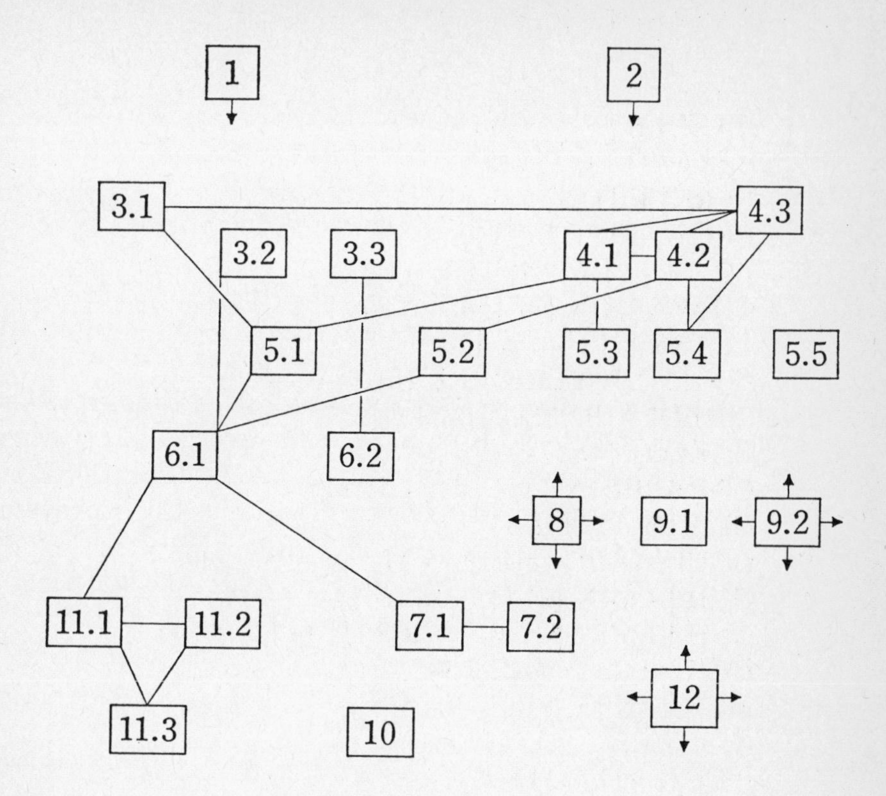

## Liste Ouvrages de Base

*Tous les paragraphes*

BERGER-GAUDUCHON-MAZET [1]
SIMON-WISSNER [1, 2]
YAU [3, 4]

*Autres*

BANDLE [1, 3] (§11)
BÉRARD [3] (§4, 5)
BERGER [1, 3] (§3, 4, 5)
BROWDER [1] (§1)
BUSER [9] (§9.1, 11.1, 11.3)
CHAZARAIN [2] (§4.2, 5.2)
CLARK [1] (§1, 2, 4)
COLIN DE VERDIÈRE [6] (§4.1, 5.1)
FRIEDLAND [2] (§1, 11)
GALLOT [4] (§11)
GARABEDIAN [1] (§1, 7.1, 11)
GILKEY [3] (§5.1)
GOULAOUIC [1] (§1, 4.3)
GUILLEMIN [2] (§4, 5.2, 9.1)
GUILLEMIN-STERNBERG [1] (§2.2, 4.5)
HEJHAL [1] (§4.3, 9.1)
Mc KEAN [3] (§6.1)
OSSERMAN [2] (§11)
PAYNE [1] (§11)
POLYA-SZEGO [1] (§11)
PROTTER [3] (§1, 2.1, 11)
REED-SIMON [1] (§1)
VENKOV [1] (§2, 4.3, 5.3, 9.1)

## Liste Chronologique des Colloques Cités en Références

1973 CIME 1973; Proceedings of Symposia n° 27 A.M.S.
1977 Partial differential equations and geometry Stochastic Differential Equations and Applications
Minimal submanifolds including geodesics
1978 Global Analysis; Pseudo-differential Operators with Applications; Probabilistic analysis and related topics
1979 Geometry of the Laplace operator; Non linear problems in Geometry
1980 Geometry and analysis; Free Boundary Problems I & II; Geometry Symposium;

1981 Séminaire Franco-Japonais; Contribution to Analysis and Geometry; Global Differential Geometry and Global Analysis; Nonlinear Partial Differential Equations and their Applications; Seminar on Harmonic Analysis; Spectral Theory of Differential Operators;

1982 Differential Geometry; Differential Geometric Methods in Mathmatical Physics; Seminar on Differential Geometry;

## Références


Albert, J. H.

1. Topology of the nodal and critical sets for eigenfunctions of elliptic operators, Thèse M.I.T. 1971.
2. Nodal and critical sets for eigenfunctions of elliptic operators, in Proc. Symp. in Pure Math, vol. 23, A.M.S. 1973, p. 71–78.
3. Genericity of simple eigenvalues for elliptic PDE'S, Proc. Amer. Math. Soc. **48** (1975), 413–418.
4. Generic properties of eigenfunctions of elliptic partial differential operators, Trans. Amer. Math. Soc. **238** (1978), 341–354.

Alimov-Il'in-Mikishin

1. Convergence problems of multiple trigonometric series and spectral decompositions I & II, Russian Math. Surveys **31** (1976), 29–86 & **32** (1977), 115–139.

Aomoto, K.

1. Formule variationnelle de Hadamard et modèles des variétés différentiables plongées, J. Funct. Analysis **34** (1979), 493–523.

Arnal, D.

1. Solutions de l'équation de la chaleur et vecteurs analytiques du Laplacien dans une représentation d'un groupe de Lie, C. R. Acad. Sci. Paris **291** (1980), 431–433.

Arnold, J. M.

1. An asymptotic theory of clad inhomogeneous planar wave guides I: Eigenfunctions and eigenvalue equations, J. Physics A **13** (1980), 3037–3081; II: Solutions of eigenvalue equations, ibid. 3083–3095.

Arnold, V. I.

1. Modes and Quasi-modes, Funct. Analysis and its Applications **6** (1972), 94–101.

Asada, S.

1. On the first eigenvalue of the Laplacian acting on p-forms, Hokkaido Math. J. **9** (1980), 112–122.
2. Notes on eigenvalues of Laplacians acting on p-forms, Hokkaido Math. J. **8** (1979), 220–227.

Asakura, F.

1. The asymptotic distribution of eigenvalues for the Laplacian in semi-infinite domains, J. of Math. of Kyoto U. **19** (1979), 583–599.

Asurov, R. R.

1. The asymptotic behavior of spectral functions of some elliptic operators (Russian) L, Differential nye Uravnenija **18** (1982), 621–625, 733.

Atiyah, M. F.

1. Classical groups and classical differential operators on manifolds, in CIME 1975:3 (Varenna), p. 5–48.

Atiyah, M.-Bott, R.-Patodi, V. K.

1. On the heat equation and the index theorem, Inventiones Math. **19** (1973),

279–330.
Atiyah, M.-Donnelly, H.-Singer, I.
1. Geometry and analysis of Shimizu L-functions, Proc. Natl. Acad. Sci. U.S.A. **79** (1982), 5751.
Atiyah, M.-Patodi, V. K.-Singer, I. M.
1. Spectral asymmetry and Riemannian Geometry I, II, III, Math. Proc. Cambridge Phil. Soc. **77** (1975), 43–69, **78** (1975), 405–432 & **79** (1976), 71–99.
Aubin, T.
1. Fonctions de Green et valeurs propres du Laplacien, J. Math. Pures et appl. **53** (1974), 347–371.
2. Meilleures constantes dans le théorème d'inclusion de Sobolev et un théorème de Fredholm non linéaire pour la transformation conforme de la courbure scalaire, J. Funct. Analysis **32** (1979), 148–174.
3. Inégalités concernant la première valeur propre non nulle du Laplacien pour certaines variétés riemanniennes, C.R.A.S. **281** (1975), 979–982.
4. Problèmes isopérimétriques et espaces de Sobolev, J. Diff. Geom. **11** (1976), 573–598.
Babich, V. M.
1. The asymptotic behaviour of "quasi-eigenvalues" of the exterior problem for the Laplace operator, in Topics in Math. Physics, 2. edited by M. Sh. Birman, Plenum Press, New York, 1968.
Babich, V. M.-Lazutkin, V. F.
1. Eigenfunctions concentrated near a closed geodesic, in Topics in Math. Physics, 2. edited by M. Sh. Birman, Plenum Press, New York 1968, p. 9–18.
Babich, V. M.-Levitan, B. M.
1. The focusing problem and asymptotic behaviour of the spectral function of the Laplace-Beltrami operator I (Russian-English summary) Math. Questions in the theory of wave propagation, 9. Zap. Naučn. Sem. Leningrad Otdel. Mat. Inst. Steklov (LOMI) **78** (1978), 3–19, 246 & Soviet Math. Dokl. **17** (1976), 1414–1417.
Babic, V. M.-Ulin, V. V.
1. Complex ray solutions and eigenfunctions that are concentrated in the neighborhood of a closed geodesic (Russian), Zap. Naučn. Sem. Leningrad. Otdel. Mat. Inst. Steklov. (LOMI) **104** (1981), 6–13, 235.
Baider, A.
1. Non compact Riemannian manifolds with discrete spectra, J. Diff. Geom. **14** (1979), 41–57.
Balian, R.-Bloch, C.
1. Distribution of eigenfrequencies for the wave equation in a finite domain, I: three dimensional problem with smooth boundary surface, Annals of Physics **60** (1970) ,401–447.
II: Electromagnetic field, Riemannian spaces, Ibid. **64** (1971), 271–307.
III: Eigenfrequency density oscillations, Ibid. **69** (1972), 76–160.
2. Asymptotic evaluation of the Green function for large quantum numbers, Annals of Physics, **63** (1971), 592–606.
3. Solutions of the Schrödinger equation in terms of classical paths, Annals of Physics, **85** (1974), 514–545 & Scientific Work of Claude Bloch, North Holland.
Bandle, C.
1. Isoperimetric inequalities and applications, Monographs and Studies in Math. n° 7 Pitman 1980.
2. Extension d'une inégalité géométrique d'Alexandrov à un problème de valeurs propres et à un problème Poisson, C.R. Acad. Sci. **277** (1973), 987.
3. Isoperimetric inequalities, preprint Université de Bâle, 1982.
Bando J.-Urakawa, H.
1. Generic properties of eigenvalues of the Laplacian for compact Riemannian

manifolds, Tôhoku Math. J.
Barbosa J. L.-do Carmo, M.
1. Stability of minimal surfaces and eigenvalues of the Laplacian, in Minimal submanifolds including geodesics, ed. by M. Obata, North Holland 1979, US-Japan Seminar Tokyo 1977.
2. Stability of minimal surfaces and eigenvalues of the Laplacian, Math. Z. **173** (1980), 13–28.
3. A proof of a general isoperimetric inequality for surfaces, Math. Z. **162** (1978), 245–261.
Bardos, C.-Guillot, J. C.-Ralston, J.
1. Relation de Poisson pour l'équation des ondes dans un ouvert non borné, C.R.A.S. **290** (1980), 495–498.
2. Asymptotic expansion of the eigenvalues of the Laplacian in a bounded domain and of the eigenmodes of the wave equations in the exterior of a compact obstacle. Non-linear partial differential equations and their applications. Collège de France seminar **2** (Paris, 1979/1980) 48–63. Res. Notes in Math, 60, Pitman, 1982.
Bareket, M.
1. On the convexity of the sum of the first eigenvalues of operators depending on a real parameter, Z. Angew. Math. Phys. **32** (1981), 464–469.
Barthel, D.-Kümritz, R.
1. Laplacian with a potential, in Global Differential Geometry and Global Analysis, Proc. Berlin 1979, ed. by D. Ferrus, W. Kühnel, U. Simon & B. Wegner, Lect. Notes in Math. n° 838, Springer 1981.
Bassotti Rizza, L.
1. Numerical verification of the classical symptotic formula of H. Weyl by means of the eigenvalues of elasticity, Atti Accad Naz. Lincei. Rend. cl. Sci. Fis. Mat. Natur. **65** (1978), 171–175.
Baxley, J. V.
1. Some partial differential operators with discrete spectra, (Birmingham, Ala. 1981) pp. 53–59, North-Holland Math. Studies, 55 North-Holland, 1981.
Bedford, E.-Suwa, T.
1. Eigenvalues of Hopf manifolds, Proc. Amer. Math. Soc. **60** (1975), 259–264.
Beers, B. L.-Millman, R. S.
1. The spectra of the Laplace-Beltrami operator on compact, semi-simple Lie groups, Amer. J. of Math. **99** (1977), 801–807.
Benabdallah
1. Noyau de diffusion sur les espaces homogènes compacts, Bull. Soc. Math. France **101** (1973), 265–283.
Benko, K.-Kothe, M.-Semmler, K. D.-Simon, U.
1. Eigenvalues of the Laplacian and curvature, Colloq. Math. **42** (1978), 19–31.
Bérard, P. H.
1. Quelques remarques sur les surfaces de révolution dans $R^3$, C.R. Acad. Sciences **282** (1976), 159–161.
2. On the wave equation on a compact Riemannian manifold without conjugate points, Math. Z. **155** (1977), 249–276.
3. Heat and wave operators on compact Riemannian manifolds, Notas de Curso n° 13, Universidade Federal de Pernambuco, Recife, Brasil, 1978.
4. Spectres et groupes cristallographiques I: Domaines Euclidiens, Inventiones Math. **58** (1980), 179–199.
5. Riezs means on Riemannian manifolds, in Proc. Symp. Pure Math. n° 36, A.M.S. 1980, p. 1–12 (Geometry of the Laplace Operator).
6. Remarques sur la conjecture de Weyl, Compositio Math. **48** (1983), 35–53.
Bérard, P. H.-Besson, G.

1. Remarques sur un article de Marcel Berger: sur une inégalité pour la première valeur propre du Laplacien, Bull. Soc. Math. France **108** (1980), 333–336.
2. Spectres et groupes cristallographiques II: Domaines sphériques, Ann. Inst. Fourier, **30** (1980), 237–248.

Bérard, P. H.-Meyer, D.
1. Une généralisation de l'inégalité de Faber-Krakhn, C.R.A.S., **292** (1981), 437–439.
2. Inégalités isopérimétrique et applications, Ann. Sc. Ec. Norm. Sup. **15** (1982), 513–542.

Bérard Bergery, L.
1. Laplacien et géodésiques fermées sur les formes d'espaces hyperboliques compactes, Sém. Bourbaki (1971–1972) exposé n° 406.

Bérard Bergery, L.-Bourguignon, J. P.
1. Laplacians and Riemannian submersions with totally geodesic fibers, summary in Global Differential Geometry and Analysis proceedings Berlin 1979, Springer L. N. in Math. N° 838 p. 30–35.
2. Laplacians and Riemannian submersions with totally geodesic fibres, Illinois J. Math. **26** (1982), 181–200.

Berenstein, C. A.
1. An inverse spectral theorem and its relation to the Pompeiu problem, J. d'Analyse Math. **37** (1980), 128–144.

Berenstein, C. A.-Yang, P.
1. An overdetermined Neumann problem in the unit disk, Technical Report, U. of Maryland 1980.

Berenstein, C. A.-Zalcman, L.
1. Pompeiu's problem on symmetric spaces, Comment. Math. Helv. **55** (1980), 593–621.
2. Pompeiu's problem on spaces of constant curvature, J. Analyse Math. **30** (1976), 113–130.

Berezin, T.
1. Laplace operators on semi-simple Lie groups, Trudy Moskov Mat. Oboc **6** (1957), 371–463.

Berger, M.
1. Eigenvalues of the Laplacian, in Proc. Symp. Pure Math. n° 16, Amer. Math. Soc. 1970, p. 121–125.
2. Sur les premières valeurs propres des variétés riemanniennes, Compositio Math. **26** (1973), 129–149.
3. Geometry of the spectrum, I, in Proc. Symposia in Pure Math. n° 27, Amer. Math. Soc. 1975, p. 129–152.
4. Une inégalité universelle pour la première valeur propre du Laplacien, Bull. Soc. Math. France, **107** (1979), 3–9.
5. Some relations between volume, injectivity radius and convexity radius in Riemannian manifolds, in Diff. Geom. Relativity in honour of A. Lichnerowicz' 60th Birthday 1976, p. 33–42 ed. by Cahen-Flato, D. Reidel Publ.
6. Aire des disques et rayon d'injectivité dans les surfaces riemanniennes C.R. A.S. **292** (1981), 291–293.
7. Isosystolic and isembolic inequalities, Preprint University Paris 7, 1981.
8. Riemannian manifolds whose Ricci curvature is bounded from below, Lectures written by S. Tsujishita, Osaka U. 1982 (in Japanese).

Berger, M.-Gauduchon, P.-Mazet, E.
1. Le spectre d'une variété riemannienne, Lecture notes in Math. N° 194 Springer 1971.

Berger, M.-Kazdan, J.
1. A Sturm-Liouville inequality with applications to an isoperimetric inequality for volume in terms of injectivity radius, and to Wiedersehen manifolds,

in General Inequalities 2, ed. by E. F. Beckenbach, Birkhäuser, Basel 1980.
Berry, J. P.
1. Tores isospectraux en dimension 3, C.R.A.S., **292** (1981), 163–166.
Berthier, A. M.
1. Sur le spectre ponctuel de l'opérateur de Schrödinger, C.R.A.S. **290** (1980), 393–395.
Besse, L. Arthur
1. Manifolds all of whose geodesics are closed, Ergebnisse der Mathematik n° 93, Springer 1978.
Besson, G.
1. Sur la multiplicité de la première valeur propre des variétés riemanniennes, Annales Inst. Fourier **30** (1980), 109–128.
2. Inégalités isopérimétriques et des applications. I. (ces actes).
Bezjaev, V. I.
1. Asymptotics of the eigenvalues of hypoelliptic operators on a closed manifold, Dokl. Akad. Nauk. SSSR **244** (1979), 1054–1057.
Biollay, Y.
1. Problèmes de Sturm-Liouville: bornes pour les valeurs propres et les zéros des fonctions propres, Journal of Applied Math. and Physics (ZAMP) **24** (1973), 525–536, 730–746 and 811–885.
Bleecker, D.
1. The spectrum of a Riemannian manifold with a unit Killing vector field, preprint U. of Hawaii 1982.
Bleecker, D. D.-Weiner, J. L.
1. Extrinsic bounds on $\lambda_1$ of $\Delta$ on a compact manifold, Commentarii Math. Helv. **51** (1976), 601–609.
Bleecker, D. D.-Wilson, L. C.
1. Splitting the spectrum of a Riemannian manifold, Siam J. Math. Analysis **11** (1980), 813–818.
Boimatov, K. A.-Kostjučenko, A. G.
1. The asymptotic behavior of Riez means of the spectral function of an elliptic operator, Dokl. Akad. Nauk. SSSR **241** (1978), 517–520.
Bonami, A.-Clerc, J. L.
1. Sommes de Cesàro et multiplicateurs des développements en harmoniques sphériques, Trans. Amer. Math. Soc. **183** (1973), 223–263.
Borell, C.
1. Convex measures on locally convex spaces, Arkiv för Mat. **12** (1974), 239–252.
Borg, G.
1. Eine Umkehrung der Sturm-Liouvilleschen Aufgabe, Acta Math. **78** (1945), 1–96.
Bourguignon, J. P.
1. Première valeur propre du Laplacien et volume des sphères riemanniennes, in Séminaire Goulaouic-Schwartz 1979–1980, Ecole Polytechnique.
Boutet de Monvel, L.
1. Nombre de valeurs propres d'un opérateur elliptique et polynôme de Hilbert-Samuel, Séminaire Bourbaki 1978–1979, exposé n° 532.
2. Convergence dans le domaine complexe de séries de fonctions propres, C.R.A.S. **287** (1978), 855–856 & Journées SMF, Eq. Deri. Part. Mai 1979
3. Opérateurs à coefficients polynomiaux, espaces de Bargman et opérateurs de Toeplitz, in Séminaire Goulaouic-Meyer-Schwartz 1980–1981, Ecole polytechnique.
Boutet de Monvel, L.-Grivard, P.
1. Le comportement asymptotique des valeurs propres d'un opérateur, C.R.A.S. **272** (1971), 23–26.
Boutet de Monvel, L.-Guillemin, V. W.

1. The spectral theory of Toeplitz operators, Annals of Math. Studies n° 99, Princeton U. Press 1981.

Brascamp, H. J.-Lieb, E. H.

1. A logarithmic concavity theorem with some applications, J. Functional Analysis **22** (1976), 366–389.
2. Some inequalities for Gaussian measures and the long range order of the one dimensional plasma, in Functional integration and its applications, edited by A. M. Arthurs, Clarendon Oxford 1975.
3. On extensions of the Brunn-Minkovski and Prékopaleindler theorems, including inequalities for log concave functions and with an application to the diffusion equation, J. Funct. Analysis **22** (1976), 366–389.

Brooks, R.

1. Exponential growth and the spectrum of the Laplacian, Proc. Amer. Math. Soc. **82** (1981), 473–477.
2. A relation between growth and the spectrum of the Laplacian, Math. Z. **178** (1981) 501–508.
3. The fundamental group and the spectrum of the Laplacian, Comment. Math. Helv. **56** (1981), 581–598.
4. Amenability and the spectrum of the Laplacian, Bull. Amer. Math. Soc. **6** (1982) 87–89.
5. The spectral geometry of foliations, preprint, Univ. of Maryland, 1981.

Browder, F. E.

1. On the spectral theory of elliptic differential operators I, Math. Ann., **142** (1961), 22–130; II, Math. Ann. **145** (1962).

Brüning, J.

1. Zur Abschätzung der Spektralfunktion elliptischer Operatoren, Math. Z. **137** (1974), 75–85.
2. Über Knoten von Eigenfunktionen des Laplace-Beltrami operators, Math. Z. **158** (1978), 15–21.
3. Invariant eigenfunctions of the Laplacian and their asymptotic distribution, in Global Differential Geometry and Analysis, Proceedings Berlin 1979, Springer L. N. in Math. n° 838, p. 69–81.
4. Über Membranen mit speziellen Knotenlinien, Comment. Math. Helv. **55** (1980), 13–19.

Brüning, J.-Gromes, D.

1. Über die Länge der Knotenlienien schwingender Membranen, Math. Z. **124** (1972), 79–82.

Brüning, J.-Heintze, E.

1. Representations of compact Lie groups and elliptic operators, Inventiones Math. **50** (1979), 169–203.

Buser, P.

1. Riemannsche Flächen mit Eigenwerten in (0, 1/4), Commentarii Math. Helv. **52** (1977), 25–34.
2. Ein untere Schranke für $\lambda_1$ auf Mannigfaltigkeiten mit fast negativer Krümmung, Archiv der Math. **30** (1978), 528–531.
3. Riemannsche Flächen mit grosser Kragenwerte, Commentarii Math. Helv. **53** (1978), 395–407.
4. Über eine Ungleichung von Cheeger, Math. Z. **158** (1978), 245–252.
5. Cubic graphs and the first eigenvalue of a Riemann surface, Math Z. **162** (1978), 87–99.
6. Über den ersten Eigenwert des Laplace-Operators auf kompakten Mannigfaltigkeiten nicht positiver Krümmung, Comment. Math. Helv. **54** (1979), 477–493.
7. Dichtepunkte in Spektrum Riemannscher Flächen, ibid, 431–439.
8. Beispiele für $\lambda_1$ auf kompakten Mannigfaltigkeiten, Math. Z. **165** (1979), 107–133.

9. On Cheeger Inequality $\lambda_1 \geq h^2/4$, in Proc. Symp. Pure Math n° 36, Amer. Math. Soc. 1980 p. 29–77 (Geometry of the Laplace Operator).
10. Riemannsche Flächen und Längenspektrum von trigonometrischen Standtpunkt aus, Thèse Bonn. 1980.
11. Sur le spectre des longueurs des surfaces de Riemann, C.R.A.S. (1980).
12. A note on the isoperimetric constant, Ann. Sc. Ec. Norm. Sup.
13. The collar theorem and examples, Manuscripta Math. **25** (1978), 349–357.
14. Untersuchungen über den ersten Eigenwert des Laplace Operators auf Kompakten Flächen, Thèse Bâle 1976.

Buzzanca, C.
1. Le Laplacien de Lichnerowicz sur les surfaces à courbure négative et constante, C.R.A.S. **285** (1977), 391–393.

Cahn, R. S.
1. The asymptotic expansion of the zeta function of a compact semi-simple Lie group. Proc. Amer. Math. Soc. **54** (1976), 459–452.

Cahn, R. S.-Gilkey, P. B.-Wolf, J. A.
1. Heat equation, proportionality principle, and volume of fundamental domains, in Differential Geometry and Relativity, ed. by Cahen-Flato, D. Reidel 1976.

Cahn, R. S.-Wolf, J. A.
1. Zeta functions and their asymptotic expansions for compact symmetric spaces of rank one, Commentarii Math. Helv. **51** (1976), 1–21.

Calogero, F.
1. Isospectral matrices and polynomials, Nuovo Cimento B. **58** (1980), 169–180.
2. Isospectral matrices and classical polynomials. Linear Algebra Appl. **44** (1982) 55–60.

Campi, S.
1. On the reconstruction of a function on a sphere by its integrals over great circles, Boll. Un. Mat. Ital. **18** (1981), 195–215.

Cantor, M.-Brill, D.
1. The Laplacian on asymptotically flat manifolds and the specification of scalar curvature, Composito Math. **43** (1981), 317–330.

Carison, R.
1. Construction of isospectral deformations of differential operators with periodic coefficients. J. Funct. Anal. **46** (1982) 265–279.

Carleman, T.
1. Über die asymptotische Verteilung der Eigenwerte partieller Differential gleichungen, Berichte ü.d. Verhandlungen Sächsischen Akad. d. Wissenschaften zu Leipzig **88** (1936), 119–132.

do Carmo, M. P.
1. Stability of minimal submanifolds, in Global Differential Geometry and Global Analysis, Proc. Berlin 1979, ed. by D. Ferus and al., Lect. Note n° 838, Springer 1981.

Carroll, R.
1. A survey of some recent results in transmutation, in Spectral theory of differential operators (Birmingham, Ala, 1980 pp. 81–92, North-Holland Math. Studies, 55, North-Holland 1981.

Carroll, R.-Gilbert, J. E.
1. Some remarks on transmutation scattering theory, and special functions. Math. Ann. **258** (1981) 39–54.

Carroll, R.-Santosa, F.
1. Scattering techniques for a one dimensional inverse problem in geophysics, Math. Methods Appl. Sci. **3** (1981) 145–171.

Chachere, G.
1. Numerical experiments concerning the eigenvalues of the Laplacian on a

Zoll surface, J. Differential Geometry **15** (1980), 135–160.
Charbonnel, A. M.
1. Spectre conjoint d'opérateurs pseudo differentiels qui commutent, C.R. Acad. Sci. Paris Sér. I Math. **293** (1981) 147–150.

Chavel, I.
1. Lowest eigenvalues inequalities, in Proc. Symp. Pure Math n° 36, AMS 1980, p. 79–89 (Geometry of the Laplace operator).

Chavel, I.-Feldman, E. A.
1. The first eigenvalue of the Laplacian on manifolds of non-negative curvature, Compositio Math. **29** (1974), 43–53.
2. An optimal Poincaré inequality for convex domains of non negative curvature, Archive for Rat. Mech. Analysis **65** (1977), 263–273.
3. Spectra of domains in compact manifolds, Journal of Funct. Analysis **30** (1978), 198–222.
4. Spectra of manifolds with small handles, Comment. Math. Helv. **56** (1981), 83–102.
5. Isoperimetric inequalities on curved surfaces, Adv. in Math. **37** (1980), 83–98.
6. Isoperimetric constants of manifolds with small handles, preprint, CUNY 81.

Chazarain, J.
1. Formule de Poisson pour les variétés riemanniennes, Inventiones Math. **24** (1974), 65–82.
2. Spectre des opérateurs elliptiques et flots hamiltoniens, Séminaire Bourbaki 1974–1975.
3. Spectre d'un hamiltonien quantique et périodes des trajectoires classiques, C.R.A.S. **288** (1979), 725–728.
4. Comportement du spectre d'un hamiltonien quantique, C.R. Acad. Sci. Paris **288** (1979), 895–897.
5. Spectre d'un hamiltonien quantique et mécanique classique, Comm. Part. Diff. Eq. **5** (1980), 595–644.

Cheeger, J.
1. Analytic Torsion and the heat eqeuation, Annals of Math. **109** (1979), 259–322.
2. On the Hodge theory of Riemannian pseudo-manifolds, in Proc. of Symp. in Pure Math. n° 36, AMS 1980, p. 91–146 (Geometry of the Laplace operator).
3. A lower bound for the smallest eigenvalue of the Laplacian in Problems in Analysis, A symposium in honour of S. Bochner, Princeton U. Press, Princeton 1970, p. 195–199.
4. Spectral geometry of spaces with cone-like singularities, Preprint CUNY Stony Brook 1978.
5. The relation between the Laplacian and the diameter . . . Archiv. der Math., **19** (1968), 558–560.

Cheeger, J.-Gromov, M. Taylor, M.
1. Finite propagation speed, kernel estimates for functions of the Laplace operator, and the geometry of complete Riemannian manifolds, J. Differential Geom. **17** (1982) 15–53.

Cheeger, J.-Taylor, M.
1. On the diffraction of waves by conical singularities, I, Comm. Pure Appl. Math. **35** (1982), 275–331; II, ibid. 487–529.

Cheeger, J.-Yau, S. T.
1. A lower bound for the heat kernel, Comm. Pure and Applied Math. **34** (1981), 465–480.

Chen, B.-Y.
1. Conformal mappings and the first eigenvalue of the Laplacian, preprint

Michigan State U.
Chen, B. Y.-Vanhecke, L.
1. The spectrum of the Laplacian of Kähler manifolds, Proc. Amer. Math. Soc. **79** (1980), 82–86.
Chen, S.-S.
1. Spectra of discrete uniform subgroups of semi-simple Lie groups, Math. Ann. **237** (1978), 157–159.
Cheng, S.
1. Isoperimetric eigenvalue problem of even order differential equations, Pacific J. Math. **99** (1982), 303–315.
Cheng, S. Y.
1. Eigenfunctions and nodal sets, Commentarii Math. Helv. **51** (1976), 43–55.
2. A characterization of the 2-shere by eigenfunctions Proc. A.M.S. **55** (1976), 379–381.
3. Eigenvalue comparison theorems and its geometric applications, Math. Z. **143** (1975), 289–297.
4. Eigenfunctions and eigenvalues of the Laplacian, preprint U.C. Berkeley.
Cheng, S. Y.-Li, P.
1. Heat kernel estimates and lower bounds of eigenvalues, Comment. Math. Helv. **56** (1981), 327–338.
Cheng, S. Y.-Li P.-Yau, S. T.
1. Heat equation on minimal submanifolds and their applications, Amer. J. Math.
2. On the upper estimate of the heat kernel of a complete Riemannian manifold, Amer. J. of Math. **103** (1981), 1021–1063.
Cheng, S. Y.-Yau, S. T.
1. Differential equations on Riemannian manifolds and their geometric applications, Comm. Pure Applied Math. **28** (1975), 333–354.
Chicco, M.
1. Generalizations of the maximum principle and majorization of the solutions for elliptic operators of variational type. (Italian. English summary) Boll. Un. Math. Ital. **18** (1981), 419–456.
Chiti, G.
1. An isoperimetric inequality for the eigenfunctions of linear second order elliptic operators. Boll. Un. Mat. Ital. **1** (1982), 145–151.
Chudnosvsky, D. V.-Chudnovsky, G. V.
1. Bäcklund transformations connecting different isospectral deformation equations. Math. Phys. **22** (1981), 2518–2522.
Chung, K. L.-Li, P.
1. Comparison of probability and eigenvalue methods for the Schrödinger equation, Advances in Math.
C.I.M.E.
1. 3° ciclo 1973 Spectral Analysis.
Clark, C.
1. The asymptotic distribution of eigenvalues and eigenfrequencies for elliptic boundary value problems, Siam Review **9** (1967).
Clements, D. L.
1. Boundary value problems governed by second order elliptic systems. Monographs and Studies in Mathematics, 12. Pitman (Advanced Publishing Program) 1981.
Clerc, J.-L.
1. Sommes de Riesz et multiplicateurs sur un groupe de Lie compact, Annales Inst. Fourier, **24** (1974), 149–172.
2. Localisation des sommes de Riesz sur un groupe de Lie compact, Studia Mathematica, **55** (1976), 21–26.
3. Multipliers on symmetric spaces, in Proc. of Symposia in Pure Math. vol.

35, Amer. Math. Soc. Providence 1979.

Colin de Verdière, Y.

1. Spectre du Laplacien et longueurs des géodésiques périodiques I, II, Compositio Math. **27** (1973), 83–106, 159–184.
2. Quasi-modes sur les variétés riemanniennes, Inventiones Math. **43** (1977), 15–52.
3. Spectre conjoint d'opérateurs pseudo-différentiels qui commutent: I, le cas non intégrable, Duke Math. J. **46** (1979), 169–182.
4. Sur le spectre des opérateurs elliptiques à bicaractéristiques toutes périodiques, Commentarii Math. Helv. **54** (1979), 508–522.
5. Une formule de traces pour l'opérateur de Schrödinger dans $R^3$, Ann. Scient. Ecole Norm. Sup. **14** (1981), 27–39.
6. Propriétés asymptotiques de l'équation de la chaleur sur une variété compacte, Séminaire Bourbaki 1973–1974 Exposé n° 439.
7. Parametrix de l'équation des ondes et intégrales sur l'espace des chemins, Séminaire Goulaouic-Lions-Schwartz 1974–1975, Exposé n° 20, Ecole Polytechnique, Palaiseau.
8. Spectre conjoint d'opérateurs pseudo-différentiels qui commutent II, le cas intégrable, Math. Z. **171** (1980), 51–73.
9. La matrice de scattering pour l'opérateur de Schrödinger sur la droite réelle. Bourbaki Seminar vol. 1979/80 pp. 246–256. Lecture Notes in Math. 812, Springer, 1981.
10. Pseudo-Laplaciens I & II, preprint, Université de Grenoble, 1981.

Colin de Verdière, Y.-Frisch, M.

1. Régularité Lipschitzienne et solutions de l'équation des ondes sur une variété riemannienne compacte, Ann. Scient. Ec. Norm. Sup. **9** (1976), 539–565.

Combes, J.-M. & Ghez, J.-M.

1. Estimations a priori avec poids pour certaines perturbations non symétriques du Laplacien sur $R^n$. C.R. Acad. Sci. Paris. Sér. I Math. **294** (1982) 521–524.

Contributions to Analysis and Geometry

1. Papers from the conference held at the Johns Hopkins University, Baltimore Md., April 24–25, 1980. Edited by Douglas N. Clark, G. Pecelli and Richard Sacksteder. Johns Hopkins University Press, Baltimore, Md., 1981.

Courant, R.-Hilbert, D.

1. Methods of mathematical physics, Wiley-Interscience I (1953), II (1962).

Croke, C. B.

1. Some isoperimetric inequalities and eigenvalue estimates, Ann. Scient. Ecole Norm. Sup. **13** (1981), 419–436.
2. The first eigenvalue of the Laplacian for plane domains, Proc. Amer. Math. Soc. **81** (1981), 304–305.
3. An eigenvalue pinching theorem, Invent. Math. **68** (1982), 253–256.

Danet, P.

1. Puissances fractionnaires de l'opérateur de Laplace-Beltrami sur une variété Riemannienne compacte, C.R. Acad. Sci. **283** (1976), 701–703.

Debiard, A.-Gaveau, B.-Mazet, E.

1. Théorèmes de comparaison en géométrie riemannienne, Publ. Res. Inst. Math. Sc. Kyoto U. (R.I.M.S.) **12** (1976), 391–425.

De George, D. L.

1. Length spectrum for compact locally symmetric spaces of strictly negative curvature, Ann. Sc. Ec. Norm. Sup. **10** (1977), 133–152.

Differential Geometry

1. Seminar on Differential Geometry. Papers presented at seminars held during the academic year 1979–1980. Edited by Shing Tung Yau, Annals of Mathematics Studies, 102. Princeton University Press; University of Tokyo Press,

Tokyo, 1982.
Differential Geometric Methods in Mathematical Physics
1. Proceedings of an International Conference held at the Technical University of Clausthal, Clausthal, July 23–25, 1980, Edited by Heinz-Dietrich Doebner, Stig I. Anderson and Herbert Rainer Petry. Lecture Notes in Mathematics, 905. Springer Verlag, 1982.

Dlubek, H.-Friedrich, Th.
1. Spektral eigenschaften des Dirac Operators die Fundamentallösung seines Wärmeleitungsgleichung und die Asymptotenentwicklung der Zeta-Funktion, J. Differential Geom. **15** (1980), 1–26.

Dodziuk, J.
1. Finite-difference approach to the Hodge theory of harmonic forms, Amer. J. Math. **98** (1976), 79–104.
2. Maximum principle for parabolic inequalities and the heat flow on open manifolds, preprint Math. Inst. U. Oxford 1980.
3. Eigenvalues of the Laplacian and the heat equation, Amer. Math. Monthly. **88** (1981), 686–695.
4. De Rham-Hodge theory for $L^2$-cohomology of infinite coverings, Topology, **16** (1977), 157–165.
5. Sobolev spaces of differential forms and Rham-Hodge isomorphism, J. Differential Geom. **16** (1981), 63–73.
6. Vanishing theorems for square-integrable harmonic forms. Proc. Indian Acad. Sci. Math. Sci. **90** (1981), 21–27.
7. $L_2$-harmonic forms on rotationally symmetric Riemannian manifolds, Proc. Amer. Math. Soc. **77** (1979), 395–400.
8. $L_2$-harmonic forms on complete manifolds, Ann. Math. Studies n° 102, Princeton Univ. 1982, p. 291–301.
9. Every covering of a compact Riemann surface of genus greater than one carries a nontrivial $L^2$-harmonic differential, preprint 1982.
10. Eigenvalues of the Laplacian on forms, preprint Oxford U. 1982.

Dodziuk, J.-Patodi, V. K.
1. Riemannian structures and triangulations of manifolds, J. Indian Math. Soc. **40** (1976), 1–52.

Donnelly, H.
1. Symmetric Einstein spaces and spectral geometry, Indiana U. Math. J. **24** (1974), 603–606.
2. A spectral condition determining the Kähler property, Proc. A.M.S. **47** (1975), 187–194.
3. Minakshisundaram's coefficients on Kähler manifolds, Proc. Symp. in Pure Math n° 27, AMS 1975, p. 195–203.
4. Spectral invariants of the second variation operator, Illinois J. Math. **21** (1977), 185–189.
5. Eta invariant of a fibered manifold, Topology, **15** (1976), 247–252.
6. Spectrum and the fixed point sets of isometries I, Math. Ann. **224** (1976), 161–170.
7. Eta invariants and the boundaries of hermitian manifolds, Amer. J. Math. **99** (1977), 879–900.
8. G-spaces, the asymptotic splitting of $L^2(M)$ into irreducibles, Math. Ann. **237** (1978), 23–40.
9. On the wave equation asymptotics of a compact negatively curved surface, Inventiones Math. **45** (1978), 115–137.
10. Eta invariants for G-spaces, Indiana U. Math. J. **27** (1978), 889–918.
11. Spectral geometry for certain non compact Riemannian manifolds, Math. Z. **169** (1979), 63–76.
12. On the analytic torsion and eta invariant for negatively curved manifolds, Amer. J. Math. **101** (1979), 1365–1379.

13. Asymptotic expansions for the compact quotients of properly discontinuous group actions, Illinois J. Math. **23** (1979), 485–496.
14. Spectrum and the fixed point sets of isometries III, preprint Johns Hopkins U.
15. Expansions associated to clean intersections, J. Diff. Geom. **14** (1979), 563–588.
16. Stability theorems for the continuous spectrum of a negatively curved manifold, Trans. Amer. Math. Soc. **264** (1981), 431–448.
17. Eigenvalues embedded in the continuous spectrum for negatively curved manifold, Michigan Math. J. **28** (1981), 53–62.
18. On the essential spectrum of a complete Riemannian manifold, Topology **20** (1981), 1–14.
19. The differential form spectrum of hyperbolic spaces, Manuscripta Mathematica **33** (1981), 365–385.
20. On the cuspidal spectrum for finite volume symmetric spaces, J. Differential Geom. **17** (1982), 239–253.
21. On the spectrum of towers, preprint, I.A.S. Princeton 1982.

Donnelly, H.-Li, P.
1. Pure point spectrum and negative curvature for non compact manifolds, Duke Math. J. **46** (1979), 497–503.
2. Lower bounds for the eigenvalues of negatively curved manifolds, Math. Z. **172** (1980), 29–40.
3. Lower bounds for the eigenvalues of Riemannian manifolds, Michigan Math. J. **29** (1982) 149–161.

Donnelly, H.-Patodi, V. K.
1. Spectrum and the fixed point sets of isometries, II, Topology, **16** (1977), 1–11.

Driscoll, B. H.
1. The multiplicity of the eigenvalues of a symmetric drum, ph.D. Thesis Northwestern U. 1978.

Duistermaat, J. J.-Guillemin, V. W.
1. The spectrum of positive elliptic operators and periodic bicharacteristics, Inventiones Math. **29** (1975), 39–79.

Duistermaat, J. J.-Kolk, J. A. C.-Varadarajan, V. S.
1. Spectra of compact locally symmetric manifolds of negative curvature, Inventiones Math. **52** (1979), 27–93, Erratum, ibid **54** (1979), 101.

Ehrenpreis, L.
1. An eigenvalue problem for Riemann surfaces Ann. of Math Studies **66** (1971), 131–140.

Eichhorn, J.
1. Spektraltheorie offener Riemannscher Mannigfaltigkeiten mit einer rotation symmetrischen Metrik, Math. Nachrichten. **104** (1981), 7–30.
2. Spectral properties of open Riemannian manifolds, Preprint 1980, Coll. Math.
3. Das Spektrum von $\Delta_p$ auf offenen Riemannschen Mannigfaltigkeiten mit beschränkter Schnittkrümmung und beschränkten Kern-Tensor.
4. Der de Rhamsche Isomorphiesatz in der $L_2$-Kategorie für eine Klasse offener Mannigfaltigkeiten, Math. Nachrichten **97** (1980), 7–14.
5. Semigroups and elliptic operators on manifolds. Proceedings of the Conference Topology and Measure II Part. 2 (Rostock/Warnemünde, 1977) pp. 15–18, Ernst-Moritz-Arndt Univ., Greijswald 1980.
6. Spektraltheorie offener Riemannscher Mannigfaltigkeiten mit einer rotationssymmetrischen Metrik. Math. Nachr. **104** (1981), 7–30.
7. Abschätzungen für das Spektrum vom $\Delta_p$ auf Räumen konstanter Krümmung, preprint Greifswald U. 1982.
8. Riemannsche Mannigfaltigkeiten mit einer zylinderähnlichen Endenmetrik, preprint Greifswald U. 1982.

9. Spectrum and curvature for rotationally symmetric metrics, Proc. Conf. Diff. Geometry and Applic. Karlova U. 1980.

Ejiri, N.

1. A construction of non-flat, compact irreductible Riemannian manifolds which are isospectral but not isometric, Math. Z. **168** (1979), 207–212.

Elstrodt, J.

1. Die Resolvente zum Eigenwertproblem der automorphen Formen in der hyperbolischen Ebene.
   I, Math. Ann. **203** (1973), 295–330.
   II, Math. Z. **132** (1973), 99–134.
   III, Math. Ann. **208** (1974), 99–132.
2. Die Selbergsche Spurformel für kompakte Riemannsche Flächen, Jber d. Dt. Math. Verein **83** (1981), 45–77.

Elstrodt, J.-Roelcke, W.

1. Uber das wesentliche Spektrum zum Eigenwertproblem der automorphen Formen, Manuscripta Math. **11** (1974), 391–406.

Elworthy, K. D.

1. Stochastic methods and differential geometry. Bourbaki Seminar, Vol. 1980/81 pp. 95–110, Lecture Notes in Math. **901** Springer, 1981.

Elworthy, K. D.-Truman, A.

1. Classical mechanics, the diffusion (heat) equation and the Schrödinger equation on a Riemannian manifold, J. Math. Phys. **22** (1981) 2144–2166.

Fegan, H. D.

1. The heat equation on a compact Lie group, Trans. Amer. Math. Soc., **246** (1978), 339–357.
2. The spectrum of the Laplacian for forms over a Lie group, Pacific J. Math. **90** (1980), 373–387.
3. The heat equation and modular forms, J. Diff. Geom. **13** (1978), 589–602.
4. The Laplacian with a character as a potential and the Clebsch-Gordon numbers.

Fischer, M. E.

1. On hearing thee shape of a drum, J. of Combinatorial Theory, **1** (1966), 105–125.

Flaschka, H. et al.

1. Multiphase averaging and the inverse spectral solution of Korteweg- de Vries equation, Comm. Pure Applied Math. **33** (1980), 739–784.

Fleckinger, J.

1. Estimate of the number of eigenvalues for an operator of Schrödinger type, Proc. Roy. Soc. Edinburgh Sect. A **89** (1981) 355–361.

Fleckinger-Pellé, J.

1. Répartition des valeurs propres d'opérateurs de type Schrödinger, CRAS **292** (1981), 359–362.

Forsythe, G. E.

1. Difference methods on a digital computer for Laplacian boundary value and eigenvalue problems, Comm. in Pure Appl. Math. **9** (1956), 425–434.

Free Boundary Problems

1. Vol. II (Pavia 1979) Vol. I (Pavia 1979) 1st Naz. Alta. Mat. Francesco Severi, Rome 1980.

Friedland, S.

1. Extremal eigenvalue problems defined for certain classes of functions, Archive for Rat. Mech. and Analysis, **67** (1977), 73–81.
2. Extremal eigenvalue problems, Bol. Soc. Bras. Mat. **9** (1978), 13–40.
3. Extremal eigenvalue problems defined on conformal classes of compact Riemannian manifolds, Commentarii Math. Helv. **54** (1979), 494–507.
4. Inverse eigenvalue problems, Linear Algebra App. **17** (1977), 15–51.

Friedland, S.-Hayman, W. K.

1. Eigenvalue inequalities for the Dirichlet problem on spheres and the growth of subharmonic functions, Commentarii Math. Helv. **51** (1976), 133–161.
Friedland, S.-Nowosad, P.
1. Extremal eigenvalue problems with indefinite kernels, Adv. in Math. **40** (1981), 128–154.
Friedlander, L.
1. Sur le spectre de la perturbation faible d'un opérateur auto adjoint, C.R. Acad. Sci. Paris Sér. I. Math. **293** (1981), 465–468.
Friedrich, Th.
1. Der erste Eigenwert des Dirac-Operators einer kompakten Riemannschen Mannigfaltigkeiten nicht negativer Scalar Krümmung, Math. Nachrichten **97** (1980), 117–146.
2. A remark on the first eigenvalue of the Dirac operator on 4-dimensional manifolds, Math. Nachr. **102** (1981), 53–56.
Frisch, M.
1. Croissance asymptotique des solutions de l'équation des ondes sur une variété riemannienne compacte à courbure sectionnelle négative ou nulle, Séminaire Goulaouic Schwartz 1976–1977, Exposé n° 21, Ecole Polytechnique
2. Propriétés asymptotiques des vibrations de tores, J. Math. Pures et Appl. **54** (1975), 285–304.
3. Propriétés asymptotiques des vibrations des sphères, J. Math. Pures et Appl. **55** (1976), 421–430.
Fujiwara, D.
1. A remark on the Hadamard variational Formula I, Proc. Japan Acad. **55** (1979), 180–184.
2. A remark on the Hadamard variational Formula, II, Proc. Japan Acad. Ser. A. Math. Sci. **57** (1981), 337–341.
3. Green's function and singular variation of the domain (ces actes).
Fujiwara, D.-Tanikawa, M.-Yukita
1. The spectrum of the Laplacian and boundary perturbation I, Proc. Japan Acad. **54** (1978), 87–91.
Furutani, K.
1. On eigenvalues and eigenspaces of a Laplace operator on the sphere $S^n$, TRU Math. **17** (1981) 273–283.
Gage, M.
1. Upper bounds for the first eigenvalue of the Laplace-Beltrami operator Indiana U. Math. J. **29** (1980), 897–912.
Gallot, S.
1. Equations différentielles caractéristiques de la sphère, Ann. Scient. Ecole Norm. Sup. **12** (1979), 235–267.
2. Sur quelques applications des inégalités de Sobolev à la géométrie, Preprint Université de Savoie, 1979.
3. Un théorème de pincement et une estimation de la première valeur propre du Laplacien d'un variété riemannienne, C.R.A.S. **289** (1979), 441–444.
4. Minorations sur le $\lambda_1$ des variétés riemaniennes, Séminaire Bourbaki 1980/1981 exposé n° 569.
5. Estimées de Sobolev quantitatives sur les variétés riemanniennes et applications, C.R.A.S. **292** (1981), 375–378.
6. Variétés dont le spectre ressemble à celui de la sphère, Journées S.M.F. Analyse sur les variétés Metz 1979, Astérisque, Société Math. de France 1980.
7. Inégalités isopérimétriques sur les variétés compactes sans bord, Preprint U. de Savoie Chambéry 1981.
8. Minorations sur le $\lambda_1$ des variétés riemanniennes. Bourbaki Seminar, Vol. 1980/81 pp. 132–148. Lecture Notes in Math., 901, Springer, 1981. [same as 4]

9. Sobolev inequality and some geometric applications (ces actes).
Gallot, S.-Meyer, D.
1. Opérateur de courbure et Laplacien des formes différentielles d'une variété riemannienne, J. Math. Pures et Appliquées, **54** (1975), 259–284.
Gangolli, R.
1. Asymptotic behavior of spectra of compact quotients of certain symmetric spaces, Acta Math. **121** (1968), 151–192.
2. The length spectra of some compact manifolds of negative curvature, J. Diff. Geom. **12** (1977), 403–424.
3. Zeta functions of Selberg's type for compact space forms of symmetric spaces of rank 1.
Garabedian, P. R.
1. Partial Differential Equations, John-Wiley New York (1964).
Garabedian, P. R.-Schiffer, M.
1. Convexity of domain functionals, J. Analyse Math. **2** (1952–53), 281–368.
Gasymov, M. G.-Levitan, B. M.
1. On Sturm-Liouville operators with discrete spectra.
Gehtman, M. M.
1. On the existence of surface states for nonclassical selfadjoint extensions of the Laplace operator. (Russian) Funkcional Anal. I Prilozen **16** (1982) 62-63.
Gelfand, I. M.
1. On elliptic equations, Russian Math. Surveys.
Gel'fand, I. M. Levitan, B. M.
1. On the determination of a differential equation from its spectral function, Izvest. Akad. Nauk **15** (1951) Amer. Math. Soc. Trans. **1** (1955).
Gelfand, I. M.- Yaglom, A. M.
1. Integration in function spaces and applications to quantum physics, J. Math. Physics 1 (1960), 48–69.
Geller, D.
1. The Laplacian and the Kohn Laplacian for the sphere, J. Diff. Geom. **15** (1980), 415–435.
Geometry and Analysis
1. Papers dedicated to the memory of V. K. Patodi, Indian Acad. Sci. Bangalore, Tata Institute Bombay 1980.
Geometry of the Laplace Operator
1. Ed. by R. Osserman & A. Weinstein Proc. Symp. Pure Math. n° 36, AMS 1980.
Geometry Symposium, Utrecht, 1980
1. Edited by Eduard Looijenga, Dirk Siersma and Floris Takens, Lecture Notes in Mathematics, **894** Springer Verlag, 1981.
Gilbarg, D.-Trudinger, N. S.
1. Elliptic partial differential equations of second order, Grundlehren der math. Wissenschaften 224, Springer 1977.
Gilkey, P. B.
1. Curvature and the eigenvalues of the Laplacian for elliptic complexes, Adv. in Math. **10** (1973), 344–382.
2. Curvature and the eigenvalues of the Dolbeault complex for Kähler manifolds, Adv. in Math. **11** (1973), 311–325.
3. The index theorem and the heat equation, Lect. Notes n° 4, Publish or Perish Inc. Boston 1974.
4. Spectral geometry and the Kähler condition for complex manifolds, Inventiones Math. **26** (1974), 231–258.
5. The boundary integrand in the formula for the signature and Euler characteristic of a Riemannian manifolds with boundary, Adv. in Math. **15** (1975), 334–360.

6. The spectral geometry of a Riemannian manifolds, J. Diff. Geom. **10** (1975), 601–618.
7. The spectral geometry of real and complex manifolds, in Proc. Symp. Pure Math. n° 27, AMS 1975, p. 265–280.
8. Curvature and the eigenvalues of the Dolbeault complex for hermitian manifolds, Adv. in Math. **21** (1976), 61–77.
9. The spectral geometry of symmetric spaces, Trans. A.M.S. **225** (1977), 341–353.
10. Lefschetz fixed point formulas and the heat equation, in Partial Diff. Eq. and Geometry, ed. by CI. Byrnes, Marcel Dekker 1979.
11. The residues of the local et a functions at the origin, Math. Ann. **240** (1979), 183–189.
12. Recursion relations and the asymptotic behavior of the eigenvalues of the Laplacian, Composition Math. **38** (1979), 201–240.
13. Curvature and the heat equation for the de Rham complex, in Geometry and Analysis, Papers dedicated to the memory of V. K. Patodi, Indian Acad. Sci., Bangalore (1980), P. 47–79.
14. The spectral geometry of the higher order Laplacian, Duke Math. J. **47** (1980), 511–528.
15. Spectral geometry and the Lefschetz formulas for a holomorphic isometry of an almost complex manifolds, preprint Princeton U.
16. Spectral geometry and the generalized Lefschetz fixed point formula for the De Rham and signature complexes, preprint Princeton U.
17. Local invariants of the Riemannian metric for 2-dimensional manifolds, Indiana U. Math. J. **23** (1974), 855–881.
18. Local invariants of an embedded Riemannian manifold, Annals of Math. **102** (1975), 187–203.
19. Local invariants of a pseudo-Riemannian manifold, Math. Scandinavica **36** (1975), 109–130.
20. Local invariants of real and complex Riemannian manifolds in Proc. Symposia Pure Math. N° 30, Amer. Math. Soc. 1977.
21. Smooth local invariants of a Riemannian manifold, Adv. in Math. **28** (1978), 1–10.
22. Curvature and the heat equation for the de Rham complex. Proc. Indian Acad. Sci. Math. Sci. **90** (1981) 47–79.
23. Invariance theory, the heat equation, and the Atiyah-Singer index theorem, Publish or Perish Inc. 1982.

Gilkey, P. B.-Sacks, J.

1. Spectral geometry and manifolds of constant holomorphic curvature, in Proc. Symp. Pure Math. n° 27, AMS 1975, p. 281–285.

Gilkey, P. B.-Smith, C. L.

1. The twisted index theorem for manifolds with boundary, Preprint 1982 Eugere (Oregon).
2. The eta-invariant for a class of elliptic boundary value problems.

Glazman, I. M.

1. Direct methods of the qualitive spectral analysis of singular differential operators, Israel Program of Scientific Translations, Jerusalemn 1965.

Global Analysis

1. Lect. Notes in Math. n° 755, Springer 1979 ed. by Grmela and Marsden (Calgary 1978).

Global Differential Geometry and Global Analysis

1. Proocedings of the Colloquium held at the technical university of Berlin, Berlin, November 21–24, 1979. Edited by Dirk Ferus, Wolfgang Kühnel, Udo Simon and Bernd-Wegner. Lecture Notes in Mathematics, 838. Springer Verlag, 1981.

Goldberg, S. I.-Ishihara, T.

1. Riemannian submersions commuting with the Laplacian, J. Diff. Geom. **13** (1979), 139–144.

Good, A.

1. Cusp forms and eigenfunctions of the Laplacian, Math. Ann. **255** (1981), 523–548.

Goulaouic, C.

1. Valeurs propres de problèmes aux limites irréguliers: Applications, in C.I.M.E. III Ciclo 24/08-02/09/73 Varenna; Spectral Analysis—Ed. Cremonesa Roma 1974.

Gray, A.-Pinsky, M.

1. The mean exit time from a small geodesic ball in a Riemannian manifold, Preprint U of Maryland 1982.

Greiner, P.

1. An asymptotic expansion for the heat equation, in Proc. Symp. Pure Math. n° 16, Amer. Math. Soc. 1970.
2. An asymptotic expansion for the heat equation, Arch. Rat. Mech. Analysis **41** (1971), 163–218.
3. Spherical harmonics on the Heisenberg group, Canad. Math. Bull. **23** (1980), 383–396.

Gromov, Michael

1. Paul Levy's isoperimetric inequality, preprint I.H.E.S. 1980.

Grubb, G.

1. Sur les valeurs propres des problèmes aux limites pseudo-différentiels, C.R.A.S. **286** (1978), 199–201.
2. Estimation du reste dans l'étude des valeurs propres des problèmes aux limites pseudo-différentiels auto-adjoints, CRAS **287** (1978), 1017–1020.

Guillemin, V. W.

1. Some spectral results for the Laplace operator with potential on the n-sphere, Adv. in Math. **27** (1978), 273–286.
2. Lectures on spectral theory of elliptic operators, Duke Math. J., **44** (1977), 485–517.
3. Symplectic spinors and partial differential equations, in Géometrie symplectique et physique mathématique Colloque Int. du CNRS n° 231, Aix-en-Provence, Ed du CNRS 1976, p. 217–252.
4. Some spectral results on rank one symmetric spaces, Adv. in Math. **28** (1978), 129–137. Addendum ibid p. 138–147.
5. Band asymptotics in two dimensions, Adv. in Math. **42** (1981), 248–282.
6. Spectral Theory on $S^2$: some open questions, ibid. 283–298.

Guillemin, V. W.-Kazhdan, D.

1. Some inverse spectral results for negatively curved 2-manifolds, Topology **19** (1980), 301–312.
2. Some inverse spectral results for negatively curved n-manifolds, in Proc. Symp. Pure Math. n° 36, AMS 1980 p. 153–180 (Geometry of Laplace Operator).

Guillemin, V. W.-Melrose, R.

1. An inverse spectral result for elliptic regions in $R^2$, Adv. in Math. **32** (1979), 128–148.
2. The Poisson summation formula for manifolds with boundary, Adv. in Math. **32** (1979), 204–232.

Guillemin, V. W.-Sternberg, S.

1. Geometric Asymptotics, Math. Surveys n° 14, AMS 1977.

Guillemin, V.-Uribe, A.

1. Spectral properties of a certain class of complex potentials, Preprint M.I.T. 1982.

Guillemin, V. W.-Weinstein, A.

1. Eigenvalues associated with a closed geodesic, Bull. Amer. Math. Soc. **82**

(1976), 92–94. Correction & Addendum, ibid. p. 966.
Guillopé, L.
1. Une formule de trace pour l'opérateur de Schrödinger dans $R^n$ thèse de 3 ème cycle U. de Grenoble 1981.
2. Asymptotique de la phase de diffusion pour l'opérateur de Schrödinger avec potentiel, C.R. Acad. Sci. Paris Sec. I Math **293** (1981) 601–603.
Günther, P.
1. Problèmes de réseaux dans les espaces hyperboliques, C.R. Acad. Sc. Paris **288** (1979), 49–52.
2. Gitterpunktprobleme in symmetrischen Riemannschen Raümen vom Rang 1, Math. Nachr. **94** (1980), 5–27.
3. Eine Funktionalgleichung für den Gitterrest, Math. Nachr. **76** (1977), 5–27.
4. Einige Sätze über das Volumenelement eines Riemannschen Raumes, Publicationes Math. Debrecen **7** (1960) 78–93.
5. Poisson formula and estimations for the length spectrum of compact hyperbolic space forms. Studia Sci. Math. Hungar **14** (1979) 105–123 (1982).
Günther, P.-Schimming, R.
1. Curvature and spectrum of compact Riemannian manifolds, J. Diff. Geom. **12** (1977), 599–618.
Haitov, A.
1. Distribution of eigenvalues of the Laplace operator, Dokl. Akad. Nauk. USSR **1** (1980), 12–15 (Russe).
Hall, S. W.-Štědrý, M.
1. The Rayleigh and Van der Pol wave equations and generalizations, J. Math. Anal. Appl. **76** (1980), 378–405.
Hano, J-I.
1. The complex Laplace-Beltrami operator canonically associated to a polarized Abelian variety, Manifolds and Lie Groups, Progress in Math. Birkhäuser (1981), 109–144.
Har'el, Z.
1. Curvature invariants, volume functions and spectral expansions, preprint Technion Haifa.
2. Heat kernels and volume functions, preprint Technion 1982.
Hasegawa, T.
1. A spectral invariant of a compact Riemannian manifold with boundary, Preprint Tokyo Institute of Technology 1980.
2. Spectral geometry of closed minimal submanifolds in a space form, real or complex, Kodai Math. J. **3** (1980), 224–252.
Harthong, J.
1. Les singularités des fonctions spectrales sur une variété riemannienne infiniment aplatie, in Séminaire Goulaouic-Schwartz 1979–1980, exposé n° 8, Ecole Polytechnique.
Hejhal, D. A.
1. The Selberg trace formula and the Riemann Zeta function, Duke Math. J. **43** (1976), 441–482.
2. The Selberg trace formula for PSL (2, R), Lect. Notes in Math. n° 548, Springer 1976.
3. Sur certaines séries de Dirichlet associées aux géodésiques fermées d'une surface de Riemann compacte, CRAS **294** (1982), 273–276.
Helffer, B.-Pham, T.
1. Remarque sur la conjecture de Weyl, Math. Scand. **48** (1981), 39–40.
Helffer, B.-Robert, D.
1. Comportment asymptotique précisé du spectre d'opérateurs globalement elliptiques sur $R^n$, C.R.A.S. **292** (1971), 362–366.
2. Comportement asymptotique précisé du spectre d'opérateurs globalement elliptiques dans $R^n$. Goulaouic-Meyer-Schwartz Seminar 1980–1981, Exp.

N° II Ecole Polytechnique Palaiseau, 1981.
3. Propriétés asymptotiques du spectre d'opérateurs pseudodifférentiels sur $R^n$, Comm. Partial Differential Equations **7** (1982), 795–882.
4. Etude du spectre pour un opérateur globalement elliptique dont le symbole de Weyl présente des symétries, preprint, Université de Nantes 1981.

Helgason, S.
1. Some results on eigenfunctions on symmetric spaces and eigenspace representation, Math. Scand. **41** (1977), 79–89.
2. Invariant differential operators and eigenspace representation, preprint M.I.T.
3. Eigenspaces of the Laplacian, integral representations and irreducibility, in Proc. of Symposia n° 27, Amer. Math. Soc. 1975.

Helton, J. W.
1. An operator algebra approach to partial differential equations, Propagation of singularities and spectral theory, Indiana U. Math. J. **26** (1977), 997–1018.

Hersch, J.
1. Sur les fonctions propres des membranes vibrantes couvrant un secteur symétrique de polygône régulier ou de domaines périodiques, Commentarii Math. Helv. **41** (1966–67), 222–236.
2. Transplantation harmonique, transplantation par modules et théorèmes isopérimétriques, Commentarii Math. Helv. **44** (1969), 354–366.
3. Quatre propriétés isopérimétriques des membranes sphériques homogènes, C.R.A.S. **270** (1970), 1645–1648.
4. Membranes symétriques d'égale fréquence fondamentale, J. Math. et Phys. Appl. (ZAMP), **30** (1979), 220–233.
5. Lower bounds for membrane eigenvalues, Applicable Analysis, **3** (1973), 241–245.

Hersch, J.-Monkewitz
1. Une inégalité isopérimétrique renforçant celle de ..., C.R. Acad. Sci **273** (1971) 62–64.

Hess, H.-Schrader, R.- Uhlenbock, D. A.
1. Kato's inequality and the spectral distributions of Laplacians on compact Riemannian manifolds, J. Diff. Geom. **15** (1980), 27–38.

Hile, G. N.-Protter, M. H.
1. Inequalities for eigenvalues of the Laplacian, Indiana U. Math. J. (1980), 523–538.

Hirai, T.
1. Invariant eigendistributions of Laplace operators on real simple Lie groups IV: explicit form of the character of discrete series representations for Sp(n, k), Jap J. Math **3** (1977), 1–48.

Hochstadt, H.
1. On inverse problems associated with second order differential operators, Acta Math. **119** (1967), 173–192.

Hoffman, D.
1. Remarks on a geometric constant of Yau, preprint U. of Mass at Amherst.
2. Lower bounds on the first eigenvalue of the Laplacian of Riemannian submanifolds, in Minimal submanifolds and geodesics, ed. by M. Obata, North Holland 1979.

Höppner, W.
1. Über Gruppeninvariante Randwertproblem, report 1980:2 Akademie der Wissenschaften der D.D.R. ZIMM Berlin.

Hörmander, L.
1. On the Riesz means of spectral functions and eigenfunctions expansions for elliptic differential operators, Some recent advances in basic sciences, Yeshiva U. Conference 1966, p. 155–202.

2. The spectral function of an elliptic operator, Acta. Math. **121** (1968), 193–218.
3. On the asymptotic distribution of the eigenvalues of pseudo-differential operators in $R^n$, Arkiv för Matematik **17** (1979), 297–313.

Huber, H.
1. Über den ersten Eigenwert des Laplace-Operators auf kompakten Riemannschen Flächen, Commentarii Math. Helv. **49** (1974), 251–259.
2. Über den ersten Eigenwert des Laplace-Operators auf kompakten Mannigfaltigkeiten konstanter negativer Krümmung, Archiv der Math. **26** (1975), 178–182.
3. Über die Eigenwerte des Laplace-Operators auf kompakten Riemannschen Flächen, Commentarii Math. Helv. **51** (1976), 215–231.
Über die Eigenwerte des Laplace-Operators auf kompakten Riemannschen Flächen II, Commentarii Math. Helv. **53** (1978), 458–469.
4. On the spectrum of the Laplace operator on compact Riemann surfaces, in Proc. Symp. Pure Math. n° 36, AMS 1980, p. 181–184 (Geometry of the Laplace Operator).
5. Über die Darstellungen der Automorphismengruppe einer Riemannschen Fläche in den Eigenräumen des Laplace-Operators, Commentarii Math. Helv. **52** (1977), 177–184.
6. Über die Dimenzion der Eigenräume des Laplace Operators auf Riemannschen Flächen, Commentarii Math. Helv. **55** (1980), 390–397.

Ii, K.
1. Curvature and spectrum of Riemannian manifolds, preprint Tôhoku U.

Ikeda, A.
1. On the spectrum of a Riemannian manifold of positive curvature, Osaka J. Math. **17** (1980), 75–93.
2. On lens spaces which are isospectral but not isometric, Ann. Scient. Ecole Norm. Sup. **13** (1981), 303–316.
3. On the spectrum of a Riemannian manifolds of positive constant curvature II, Osaka J. Math. **17** (1980), 691–702.
4. Isospectral problem for spherical space forms (ces actes).
5. On spherical space forms which are isospectral but not isometric, preprint 1982.

Ikeda, A.-Taniguchi, Y.
1. Spectra and eigenforms of the Laplacian on $S^n$ and $P^n$ (C), Osaka J. Math. **15** (1978), 515–546.

Ikeda, A.-Yamamoto, Y.
1. On the spectra of 3-dimensional lens spaces, Osaka J. Math. **16** (1979), 447–469.

Ilias, S.
1. Sur une inégalité de Sobolev, C.R.A.S. Paris **294** (1982), 731–734.
2. Thèse de 3$^{ème}$ cycle, Université Paris 7, 1983.

I'lin, A. M.
1. Study of the asymptotic behavior of the solutions of an elliptic boundary value problem in a domain with a small hole, Trudy Seminara Ineni I.G. Petrovskogo **6** (1981), 57–82.

I'lin, V. A.-Moiseev, E. I.
1. Estimates of anti-a priori-type that are sharp with respect to order for the eigen and associated functions of the Schrödinger operator (Russian), Differential'nye Uravenija **17** (1981), 1859–1867, 1918–1919.

Ivrii, V, Ja.
1. On the second term of the spectral asymptotics for the Laplace-Beltrami operator on manifolds with boundary and for elliptic operators acting on fiberings, Soviet Math. Dokl. **21** (1980), 300–302.
2. Second term of the spectral asymptotic expansion of the Laplace-Beltrami

operator on manifolds with boundary, Funct. Analysis and Applications, **14** (1980), 98–106.
3. The asymptotic behavior of eigenvalues for some elliptic operators acting in vector bundles over a manifold with boundary (Russian), Dokl. Akad. Nauk SSRR **258** (1981), 1045–1046.
4. On the asymptotic behavior of eigenvalues for a class of elliptic operators acting in fiber bundles over a manifold with a boundary. (Russian), Dokl. Akad. Nauk SSRR, **263** (1982), 530–531.
5. Exact spectral asymptotics for elliptic operators acting in vector bundles. (Russian) Funkcional Anal. i Prilozen **16** (1982) 30–38, 96.

Iwasaki, I.-Katase, K.
1. On the spectra of Laplace operator on $\Lambda^*(S^n)$, Proc. Japan Acad. **55** (1979), 141–145.

Jenni, F. W.
1. Über das Spektrum des Laplace-Operators auf einer Schar kompakter Riemannscher Flächen, Thesis Basel 1981.
2. Über den ersten Eigenwert des Laplace-operators auf ausgewählten Beispielen kompakter Riemannscher Flächen, preprint Basel U. 1982.

Jerison, D. S.
1. The Dirichlet problem for the Kohn Laplacian on the Heisenberg group I, J. Funct. Anal. **43** (1981), 97–142.
2. The Dirichlet problem for the Kohn Laplacian on the Heisenberg group II. J. Funct. Anal. **43** (1981), 224–257.

Jørgensen, P.
1. Spectral theory for domains in $R^n$ of finite measure, preprint Aarhus 1979.
2. Spectral theory of finite volume domains in $R^n$, preprint Aarhus 1980.
3. Point spectrum of semi-bounded operator extensions, Aahrus 1980.

Kac, M.
1. Can one hear the shape of a drum, Amer. Math. Monthly **73** (1966).
2. On applying mathematics: reflections and examples, Quaterly of Applied Math. **30** (1972).

Kalf, H.
1. Nonexistence of eigenvalues of Dirac operators, Proc. Roy. Soc. Edinburgh Sect. A **89** (1981), 309–317.

Kalka, M.-Menikoff, A.
1. The wave equation on a cone, Comm. Partial Differential equations 7 (1982), 223–278.

Kalnins, E. G.-Miller, W.
1. The wave equation and separation of variables on the complex sphere $S^4$, J. Math. Anal. Appl. **83** (1981), 449–469.

Kannai, Y.
1. Off diagonal short time asymptotics for fundamental solutions of diffusion equations, Comm. Partial diff. Eq. **2** (1977), 781–830.

Kashiwara, M.-Kowata, A.-Minemura, K.-Okamoto, K.-Oshima, T.-Tanaka, M.
1. Eigenfunctions of invariant differential operators on a symmetric space, Ann. of Math. **107** (1978), 1–39.

Kasue, A.
1. A Laplacian comparison theorem and function theoretic properties of a complete Riemannian manifold, preprint University of Tokyo 1981.
2. On Laplacian and Hessian comparison theorems, preprint 1981, Tokyo U.
3. On a lower bound for the first eigenvalue of the Laplace operator on a Riemannian manifold, preprint, 1982, Tokyo U.
4. On Riemannian manifolds with boundary, preprint Tokyo U. 1982.
5. On Laplacian and Hessian comparison theorems, Proc. Japan Acad. Ser. A Math. Sci **58** (1982), 25–28.

Keller, J. B.

1. Progress and Prospects in the theory of linear wave propagation, Siam Review **21** (1979), 229–245.

Keller, J.-Rubinow, S.

1. Asymptotic solutions of eigenvalue problems, Annals of Physics (1960), 24–75.

Kenig, C.-Tomas, P.

1. Divergence of eigenfunctions expansions, Preprint Princeton U. 1980.

Kennedy, G.

1. Some finite temperature quantum field calculations in curved manifolds with boundary, U. of Manchester.

Kitaoka, Y.

1. Positive definite quadratic forms with the same representation numbers, Archiv der Math. **28** (1977), 495–497.

Kobayashi, S.

1. Isometric imbeddings of compact symmetric spaces, Tôhoku Math. J. **20** (1968), 21-25.

Kobayashi, S.-Takeuchi, M.

1. Minimal imbeddings of R-spaces, J. Diff. Geometry **2** (1968), 203–215.

Kohler-Jobin, M. T.

1. Une méthode de comparaison isopérimétrique de fonctionnelles de domaines de la physique mathématique, Journal de Math. et de Physique Appliquées (ZAMP), **29** (1978), 757–776.
2. Démonstration de l'inégalité isopérimétrique $P\lambda^2 \geq j_0^2/2$ conjecturé par Pólya et Szegö, C. R. Acad. Scien. **281** (1975), 119–121.
3. Une inégalité isopérimétrique entre la fréquence fondamentale d'une membrane inhomogène et l'énergie d'équilibre du problème de Poisson correspondant, C.R. Acad. Scien. **283** (1976), 65–68.
4. Une propriété de monotonie isopérimétrique qui contient plusieurs théorèmes classiques, C.R. Acad. Scien. **284** (1977), 917–920.
5. Sur la première fonction propre d'une membrane: une extension à N dimensions de l'ingalité isopérimétrique de Payne-Rayner, J. Math. et Phys. Appl. (ZAMP), **28** (1977), 1137–1140.
6. Symmetrization with equal Dirichlet integrals, Siam J. Math. Anal. **13** (1982), 153–161.
7. Isoperimetric monotonicity and isoperimetric inequalities of Payne-Rayner type for the first eigenfunction of the Helmholtz problem, Z. Angew. Math. Phys. **32** (1981), 625–646.

Kolk, J. A. C.

1. Formule de Poisson et distribution asymptotique du spectre commun des opérateurs différentiels, preprint n° 46, U. Utrecht.
2. The Selberg trace formula and asymptotic behaviour of spectra, Thèse Utrecht 1977.

Komorowski, J.

1. On an estimate from below for the first positive, eigenvalue of $\Delta$, Bull. Acad. Polonaise des Sciences, **25** (1977), 999–1006.
2. A minorization of the first positive eigenvalue of the scalar Laplacian on a compact Riemannian manifold, preprint n° 92, Inst. of Math. Polish Acad. of Sciences.
3. Nets on a Riemannian manifold and finite dimensional approximation of the Laplacian, preprint n° 83, Institute of Math., Polish Academy of Sciences.

Krein, M. G.

1. On certain problems on the maximum and minimum of characteristic values and on the Lyapunov zones of stability, Amer. Math. Soc. Translations **1** (1955), 163–187.

Kudla, S. S.-Millson, J. J.

1. Harmonic differentials an closed geodesics on a Riemann surface, Inventiones Math. **54** (1979), 193–212.

Kurylev, Ya. V.

1. Asymptotics close to the boundary of a spectral function of an elliptic second order differential operator, J. Funct. Analysis **14** (1980), 236–238.

Kuwabara, R.

1. On isospectral deformations of Riemannian metrics, Compositio Math. **40** (1980), 319–324.
2. A local characterization of flat metrics and isopectral deformations, Comment. Math. Helv. **55** (1980), 427–444. & Comment. Math. Helv. **56** (1981),
3. On the spectrum of the Laplacian on vector bundles, Preprint Kyoto U. 1980.
4. On isospectral deformations of Riemannian metrics. II, Preprint Université de Tokushima, 1981.

Lange, F. J.-Simon, U.

1. Eigenvalues and eigenfunctions of Riemannian manifolds, Proceedings A.M.S. **77** (1979), 237–242.

Lascar, B.

1. Le noyau de l'équation des ondes sur une variété riemannienne compacte comme intégrale des chemins, J. d'Analyse Mathématique.

Lax, P. D.

1. The multiplicity of eigenvalues, Bull. Amer. Math. Soc. (N.S.) **6** (1982), 213–214.

Lax, P. D.-Phillips, R.

1. Scattering theory for automorphic forms, Ann. of Math. Studies n° 87, Princeton U. Press 1976.
2. Scattering theory for automorphic functions, Bull. AMS **2** (1980), 261–295.
3. The asymptotic distribution of lattice points in Euclidean and non-Eculidean spaces, J. Funct. Anal. **46** (1982) 280–350.

Lazutkin, V. F.

1. The existence of caustics for a billiard problem in a convex domain, Math. USSR Izvestija **7** (1973), 185–214.
2. Asymptotics of the eigenvalues of the Laplacian and quasi-modes...ibid, 439–466.
3. Eigenfunctions with a given caustic, USSR Compt. Math. and Math. Phys. **10** (1970), 105–120.

Lazutkin, V. F.-Terman, D. Ja.

1. On the estimate of the remainder term in a formula of H. Weyl (Russian) Funkcional, Anail. Prilozen **15** (1981) n°4 p. 81–82.
2. On the quantity of quasimodes of "bouncing ball" type (Russian), Zap. Naučn. Sem. Leningrad, Otdel Mat. Inst. Steklov (LOMI) **117** (1981), 172–182, 200.

Levinson, N.

1. The inverse Sturm-Liouville problem, Math. Tidsskr. B. (1949), 25–30.

Levitan, B. M.

1. On the determination of a Sturm-Lioville equation by two spectra.

Levy-Bruhl, A.

1. Courbure riemannienne et développements infinitésimaux du Laplacien, C.R.A.S. **279** (1974), 197–200.
2. Spectre du Laplacien de Hodge-de Rham sur les formes de degré 1 des sphères de $R^n (n \geq 6)$, Bull. Sc. Math. **99** (1975), 213–240.
3. Spectre du Laplacien de Hodge-de Rham sur $CP^n$, Bull. Sciences Math. **140** (1980), 135–143.

Li, P.

1. A lower bound for the first eigenvalue of the Laplacian on compact manifolds, Indiana U. Math. J. **28** (1979), 1013–1019.

2. On the Sobolev constant and the p-spectrum of a compact Riemannian manifold, Ann. Scient. Ecole Norm. Sup. **13** (1981), 451–457.
3. Eigenvalue estimates on homogeneous manifolds, Comment. Math Helv. **55** (1980), 347–363.
4. Eigenvalue estimates on homogeneous manifolds, preprint I.A.S., Princeton 1980. [same as 3]
5. Minimal immersions of compact irreductible homogeneous Riemannian manifolds, J. Diff. Geom. **16** (1981), 105–115.
6. Poincaré inequalities on Riemannian manifolds, Annals Math. Studies **102** Princeton U. Press. 1982, p. 73–83.

Li, P.-Treibergs, A. E.
1. Pinching theorem for the first eigenvalue on positively curved four-manifolds, Invent. Math. **66** (1982), 35–38.

Li, P.-Yau, S. T.
1. Estimates of eigenvalues of a compact Riemannian manifold, in Proc. Symp. Pure Math. n° 36, A.M.S. 1980, p. 205–239 (Geometry of the Laplace operator).
2. A conformal invariant and applications to the Willmore conjecture and the first eigenvalue for compact surfaces, Invent. Math. (1982).
3. On the Schrödinger equation and the eigenvalue problem, preprint Stanford U. 1982.

Li, P.-Zhong, J. Q.
1. Pinching theorem for the first eigenvalue on positively curved manifolds, Invent. Math. **65** (1981/82), 221–225.

Lichnerowicz, A.
1. Géométrie des groupes de transformations, Dunod, Paris 1958.

Lieb, E.
1. Bounds on the eigenvalues of the Laplace and Schrödinger operators, Bull. AMS **82** (1976), 751–753.
2. A lower bound for level spacings, Annals of physics, **103** (1977), 88.
3. The number of bound states of one-body Schrödinger operators and the Weyl problem, in Proc. Symp. Pure Math. n° 36, A.M.S. 1980, p. 241–252 (Geometry of the Laplace operator).

Lions, P. L.
1. Une inégalité pour les opérateurs elliptiques du second ordre, Ann. Math. Pura Appl. **127** (1981), 1–11.

Lobo Hidalgo, M.-Sanchez Palencia, E.
1. Sur certaines propriétés spectrales des perturbations du domaine dans les problèmes aux limites, preprint Universidad Autonoma de Madrid.

Mahar, T. J.-Willner, B. E.
1. An extremal eigenvalue problem, Comm. Pure Applied Math. **29** (1976), 517–529.

Majda, A.-Ralston, J.
1. Geometry in the scattering phase, in Proc. Symp. Pure Math. n° 36, A.M.S. 1980, p. 253–255 (Geometry of the Laplace operator).
2. An analogue of Weyl's theorem for unbounded domains I, II, III: Duke Math. J. **45** (1978), 183–196; **45** (1978), 513–536; preprint.

Malliavin, P.
1. Asymptotic of the Green function of a Riemannian manifold and Itô's stochastic integrals, Proc. Nat. Acad. Sc. U.S.A. **71** (1974).

Mallows, C. L.-Clark, J. M. C.
1. Linear-intercept distributions do not characterize plane sets, J. Appl. Prob. **7** (1970), 240–244.

Marbes, H.
1. On the spectra of $\Gamma\backslash SL(2,R)/SO(2)$ and $\Gamma\backslash SL(2,C)/SU(2)$, Math. Nachr. **104** (1981), 61–81.

2. On the spectra of compact locally symnetric Riemannian manifolds, Math. Nachr. **104** (1981), 83–99.

Marcellini, P.

1. Bounds for the third membrane eigenvalue, J. Diff. Eq. **37** (1980), 438–443.

Marvizi, S.-Melrose, R.

1. Some spectrally isolated convex planar regions, Preprint M.I.T., 1982.
2. Spectral invariants of convex planar regions, Preprint M.I.T., 1982.

Matsuzawa, T.-Tanno, S.

1. Estimates of the first eigenvalue of a big cup domain of the two-sphere, Preprint Tokyo Institute of Technology 1981.

Mazet, E.

1. Une majoration du type de Cheeger, C.R.A.S. **277** (1973), 171–174.

Maz'ja, V.G.-Nazarov, S. A.-Plamenevskii, B. A.

1. The asymptotic behavior of solutions of the Dirichlet problem in a domain with a cut out thin pipe. (Russian) Math. Sb. (N.S.) **116** (158) (1981), 187–217;
2. Asymptotic behavior of the solution of the Dirichlet problem in a domain with a thin bridge (Russian), Funktional Anal. I Prilozen. **16** (1982), 39–46, 96.
3. Asymptotics of the solutions of the Dirichlet problem in a domain with an excluded thin tube (Russian), Uspehi Mat. Nauk. **36** (1981), 183–184.

McKean, H. P.

1. An upper bound for the spectrum of $\Delta$ on a manifold of negative curvature, J. Diff. Geom. **4** (1970), 359–366.
2. Selberg's trace formula as applied to a compact Riemann surface, Comm. Pure and Appl. Math. **25** (1972), 225–246.
3. Integrable systems and algebraic curves, in Global Analysis, Lect. Notes in Math. n° 755, éd. by M. Grmela and J. E. Marschen, Springer 1979.

McKean, H. P.-Singer, I. M.

1. Curvature and the eigenvalues of the Laplacian, J. Diff. Geom. **1** (1967), 43–69.
2. Curvature and characteristic classes of compact riemannian manifolds, J. Diff. Geom. **1** (1967), 89–97.

McKean, H. P.-Van Moerbeke, P.

1. The spectrum of Hill's equation, Inventiones Math. **30** (1975), 217–274.

Meaney, C.

1. On almost-everywhere convergent eigenfunction expansions of the Laplace-Beltrami operator, preprint, 1982.

Melrose, R. B.

1. Weyl's conjecture for manifolds with concave boundary, in Proc. Symp. Pure Math. n° 36, AMS 1980, p. 257–274 (Geometry of the Laplace operator).
2. Theorème de Weyl, conjecture de Polya et formule de Lax et Phillips, Nonlinear partial differential equations and their applications, Collège de France Seminar **2** (Paris, 1979/80) 327–336, 396, Res. Notes in Math. 60, Pitman, 1982.
3. Scattering theory and the trace of the wave group. J. Funct. Anal. **45** (1982) 29–40.

Meyer, D.

1. Inégalités isopérimétriques et des applications. II, (ces actes).

Meyer, Y.

1. Trois problèmes sur les sommes trigonométriques, 1: propriétés asymptotiques des vibrations des sphères, Astérisque n° 1, S.M.F. 1973.

Miatello, R.

1. The Minakshisundaram-Pleijel coefficient for the vector-valued heat kernel on compact locally symmetric spaces of negative curvature, Trans. Amer.

Math. Soc. **260** (1980), 1–33.
Millman, R. S.
1. Remarks on the spectrum of the Laplace-Beltrami operator in the middle dimension, Tensor **34** (1980), 94–96.
Millson, J.
1. Closed geodesics and the $\eta$-invariant, Ann. of Math. **108** (1978), 1–39.
Minakshisundaram, S.-Pleijel, A.
1. Some properties of the Laplace operator on Riemannian manifolds, Canad. J. Math. **1** (1949), 242–256.
Minimal submanifolds Including Geodesics
Edited by M. Obata (US-Japan Seminar Tokyo 1977) North Holland 1979.
Mneimne, R.
1. Equation de la chaleur sur un espace riemannien symétrique et formule de Plancherel, Preprint E.N.S. Saint Cloud 1981.
Molchanov, S. A.
1. Diffusion processes and Riemannian geometry, Russian Math. Surveys **30** (1975), 1–63.
Morse-Feshbach
1. Methods of theoretical physics, 2 volumes, McGraw Hill 1953.
Moser, J.
1. Three integrable hamiltonian systems connected with isospectral deformations, Adv. in Math **16** (1975), 197–220.
2. Geometry of quadrics and spectral theory.
3. An example of a Schrödinger equation with almost periodic potential and nowhere dense spectrum, Commen. Math. Helv. **56** (1981), 198–224.
Müller, W.
1. Analytic torsion and R-torsion of Riemannian manifolds, Adv. in Math. **28** (1978), 233–305.
2. Spectral theory of non-compact Riemannian manifolds with cusps and a related trace formula, preprint I.H.E.S. 1980.
3. Spectral theory of non-compact Riemannian manifolds with cusps and a related trace formula, Report of Akademie der Wiss. DDR, Berlin 1982.
Müller Pfeiffer, E.-Stande, J.
1. Integral theorems for eigenfunctions of second order elliptic differential operators, Math. Nachr. **98** (1980), 37–47.
Muto, H.
1. The first eigenvalue of the Laplacian on even dimensional spheres, Tôhoku Math. J. **32** (1980), 427–432.
2. The multiplicities of the first eigenvalues of the Laplacian on spheres, Preprint Tokyo Institute of Technology.
3. A generalization of Hersch's inequality, preprint Tokyo Institute of Technology 1981.
Muto, H.-Urakawa, H.
1. On the least eigenvalue of the Laplacian for compact homogeneous spaces, Osaka J. Math. **17** (1980), 471–484.
Mutō, Y.
1. Riemannian submersion and the Laplace-Beltrami operator, Kodai Math. J. **1** (1978), 329–338.
2. Some eigenforms of the Laplace-Beltrami operators in a Riemannian submersion, J. Korean Math. Soc. **15** (1978), 39–57.
3. The effect of an infinitesimal deformation of the Riemannian metric on the least positive eigenvalue of the Laplacian on $S^n$, Preprint Yokohama U.
Nachman, A. I.
1. The wave equation on the Heisenberg group, Comm. Partial Differential Equations **7** (1982), 675–714.
Nagato, T.

1. On the minimum eigenvalues of the Laplacians in Riemannian manifolds, Sci. Papers College of General Education Univ. of Tokyo, **11** (1961), 177–182.

Nehari, Z.

1. On the principal frequency of a membrane, Pacific J. Math. **8** (1958), 285–293.

Nonlinear Partial Differential Equations and their Applications.
Collège de France, Séminar Vol. I (Paris 78/79), Res. Notes in Math. 53 Pitman, Boston Mass. 1981, II ibid. 60, 1982.

Non-Linear Problems in Geometry
Proc. Sixth. Int. Sym., Dir. Math., Tanigushi Found. Katata (Japan), ed. by T. Kotake & T. Ochiai, Sept. 1979.

Nooney, G. C.

1. On the vibrations of triangular membranes, Thèse Stanford U. 1954.

Obata, M.

1. Certain conditions for a Riemannian manifold to be isometric with a sphere, J. Math. Soc. Japan **14** (1962), 333–340.

Oersted, B.

1. The conformal invariance of Huyghen's principle, J. Diff, Geom. **16** (1981), 1–10.

Oliker, V. I.

1. Eigenvalues of the Laplacian and uniqueness in the Minkowski problem, J. Diff. Geom. **14** (1979), 93–98.
2. Some remarks on elliptic equations and infinitesimal deformation of submanifolds, Global differential geometry and global analysis (Berlin, 1979) pp. 211–220. Lectures Notes in Math. 838, Springer, 1981.

Olszak, Z.

1. The spectrum of the Laplacian and the curvature of Sasakian manifolds, in Global Differential Geometry and Analysis, Proceeding Berlin 1979 L. N. in Math. n° 838, Springer 1981, 221–228.
2. The spectrum of the Laplacian and the curvature of Sasakian manifolds, Global differential geometry and global analysis (Berlin, 1979) pp. 221–228 Lectures Notes in Math., 838, Springer, 1981. [same as 1]

Omori, H.

1. Construction problem of Riemannian manifold, preprint Okayama U. 1981.

Oshima, T.-Sekiguchi, J.

1. Eigenspaces of invariant differential operators on an affine symmetric space, Inventiones Math. **57** (1980), 1–81.

Osserman, R.

1. A note on Hayman's theorem on the bass note of a drum, Commentarii Math. Helv. **52** (1977), 545–555.
2. The isoperimetric inequality, Bull. A.M.S. **84** (1978), 1182–1238.
3. Bonnesen-style isoperimetric inequality, Amer. Math. Monthly **86** (1979), 1–29.
4. Isoperimetric inequalities and eigenvalues of the Laplacian, Proc. I.C.M., Helsinki 1978, 435–442.

Ozawa, S.

1. Pertubation of domains and Green kernels of heat equations I, II, III: Proc. Japan Acad. **54** (1978), 322–325; **55** (1979), 172–175; **55** (1979), 227–230.
2. The eigenvalues of the Laplacian and perturbation of boundary conditions, Proc. Japan Acad. **55** (1979), 121–124.
3. Remarks on Hadamard's variation of eigenvalues of the Laplacian, Proc. Japan Acad. **55** (1979), 328–333.
4. Singular Hadamard's variation of domains and eigenvalues of the Laplacian, Proc. Japan Acad. Ser. A. Math. Sci. **56** (1980), 306–310. II, Proc. Japan Acad. **57** (1981), 242–246.

5. Singular variations of domains and eigenvalues of the Laplacian, Duke Math. J. **48** (1981), 767–778.
6. The first eigenvalue of the Laplacian of two dimensional Riemannian manifolds, Tôhoku Math. J. **34** (1982), 7–14.
7. Hadamard's variation of Green kernels of heat equation and their traces I, Preprint U. of Tokyo 1980.
8. Surgery of domains and the Green's function of the Laplacian, Proc. Japan Acad. **56** (1980), 459–461.
9. Geometric surgery of domains and eigenvalues of the Laplacian (ces actes).
10. An asymptotic formula for the eigenvalues of the Laplacian in a domain with a small hole. Proc. Japan Acad. Ser. A Math. Sci. **58** (1982), 5–8.
11. The first eigenvalue of the Laplacian on two-dimensional Riemannian manifolds, Tôhoku Math. J., **34** (1982), 7–14.
12. Hadamard's variation of the Green kernels of heat equations and their traces. I, J. Math. Soc. Japan **34** (1982), 455–473.
13. Potential theory and eigenvalues of the Laplacian, Proc. Japan Acad. **58** (1982), 134–136.
14. An asymptotic formula for the eigenvalues of the Laplacian in a three dimensional domain with a hole, preprint 1982 & Proc. Japan Acad. **58** (1982), 5–8.
15. Electrostatic capacity and eigenvalues of the Laplacian, preprint 1982.

Pach, J.
1. On an isoperimetric problem, Studia Sci. Math. Hungar **13** (1978), 43–45.

Pansu, P.
1. Une inégalité isopérimétrique sur le groupe de Heisenberg, C.R. Acad. Sci. Paris **295** (1982), 127–130.
2. Géométrie du groupe de Heisenberg, Thése 3ème cycle, U. Paris 7, 1982.

Paquet, L.
1. Méthode de séparation des variables et calcul des spectres d'opérateurs sur les formes différentielles, Bull. Soc. Math. France **105** (1981), 85–112.

Paris, J. C. de
1. Calcul d'un majorant de $\lambda_1$ en fonction d'un minorant de la courbure de Ricci d'après Cheeger, C.R.A.S. **279** (1974), 515–517.

Parks, H.
1. Elliptic Isoperimetric problems, Indiana Univ. Math. J. **30** (1981), 937–958.

Partial Differential Equations and Geometry
Conf. held at Park City (Utah) 1977, ed. by C. I. Byrnes, Marcel Dekker 1979.

Patodi, V. K.
1. Curvature and the fundamental solutions of the heat operator, J. Indian Math. Soc. **34** (1970), 269–285.
2. Curvature and the eigenforms of the Laplace operator, J. Diff. Geom. **5** (1971), 233–249.
3. An analytical proof of Riemann-Roch-Hirzebruch theorem for Kähler manifolds, J. Diff. Geom. **5** (1971), 251–283.
4. Riemannian structures and triangulations of manifolds, in Proc. International Congress Math. 1974, p. 39–43.

Patterson, S. J.
1. The Laplace operator of a Riemann surface I, II, III: Compositio Math. **31** (1975), 83–107; **32** (1976), 71–112; **33** (1976), 227–259.

Payne, L. E.
1. Isoperimetric inequalities in mathematical physics, Siam Review **9** (1967), 453–488.
2. On two conjectures in the fixed membrane eigenvalue problem, J. Applied Math. and Phys. (ZAMP) **24** (1973), 721–729.

Payne, L. E.-Rayner, M. E.

1. An isoperimetric inequality for the first eigenfunction in the fixed membrane problem, J. of Applied Math. and Physics (ZAMP) **23** (1972), 13–15.

Peetre, J.
1. A generalization of Courant's nodal domain theorem, Math. Scand. **5** (1957), 15–20.

Perrone, D.
1. Spectrum and Lipschitz-Killing curvature in dimension 6, Rend. Sem. Mat. Univ. Politec. Torino, **38** (1980), 59–65.

Petrovsky,
1. Lectures on partial differential equations, New York, Interscience 1954.

Pham The Lai
1. Meilleures estimations asymptotiques des restes de la fonction spectrale et des valeurs propres relatifs au Laplacien, Math. Scand. **48** (1981), 5–38.

Philippin, G.
1. Comparaison entre la première valeur propre d'un domaine et celle de sa projection circulaire, J. Math. et Phys. Appl. (ZAMP) **22** (1971), 345–350.
2. Bornes inférieures pour la première valeur propre d'une plaque vibrante, C.R.A.S. (1971), 269–272.

Pinsky, M. A.
1. Stochastic Riemannian Geometry, in Probabilistic Analysis and related Topics, Vol. 1, Acad. Press 1978, ed. by
2. Spectrum of the Laplacian on a manifold of negative curvature I & II, J. Diff. Geom **13** (1978), 87–91 & ibid. **14** (1979), 609–620.
3. The eigenvalues of an equilateral triangle, Siam J. Math. Analysis **11** (1980), 819–827.
4. A topological version of Obata's sphere theorem, J. Diff. Geom. **14** (1979), 369–378.
5. An individual ergodic theorem for the diffusion on a manifold of negative curvature.
6. Stochastic Taylor formulas and Riemannian geometry, Preprint Northwestern U. 1981.
7. An individual ergodic theorem for the diffusion on a manifold of negative curvature in Stochastic Differential Equations and Applications, Acad. Press 1977, p. 231–240. [same as 5]
8. Moyenne stochastique sur une variété riemannienne. C.R. Acad. Sci. **292** (1981), 991–994.

Pleijel, A.
1. Remarks on Courant's nodal line theorem, Comm. Pure Applied Math. **9** (1956), 543–550.

Polya, G.
1. On the eigenvalues of vibrating membranes, Proc. London Math. Soc. **11** (1961), 419–433.

Polya, G.-Szegö, G.
1. Isoperimetric inequalities in mathematical physics, Annals of Math. Studies n° 27, Princeton 1951.

Probabilistic Analysis and Related Topics
1. Vol, 1, Academic Press 1978, Edited by A. T. Bharucha-Reid.

Proceedings of Symposia
In Pure Math. Vol. 27, Amer. Math. Soc. Providence N. J. 1975.

Prosser, R. T.
1. Can one see the shape of a surface, Amer. Math. Monthly.
2. Formal solutions of Inverse Scattering problems, I; II; III
J. Math. Phys. **10** (1969), 1819–1822.
**17** (1976), 1775–1779.
**21** (1980), 2648–2653.

Protter, M. H.

1. Lower bounds for the first eigenvalue of elliptic equations, Annals of Math. **71** (1960), 423–444.
2. The generalized spectrum of second order elliptic systems, Rocky Mountains J. of Math. **9** (1979), 503–518.
3. The maximum principle and eigenvalue problems, in Beijing Symposium on Diff. Geom. and Partial Diff. Eq., held summer 1980.

Pseudo Differential Operators with Applications.
1. Lectures presented at the Summar Session held by Centro Internazionale Mathematico Estivo (CIME), Bressanone, June 16–24, 1977. Liguori Editore, Naples, 1978.

Pyshkina, M. F.
1. Asymptotic hehaviour of eigenfunctions of the Helmholtz equation concentrated near a closed geodesic, in Math. Problems in wave proparation theory II, V. M. Babich ed. Sem. Math. Steklov Math. Inst. Vol. 15 Leningrad 1969—Transl. Consultants Bureau N.Y. 1971.

Ralston, J. V.
1. On the construction of quasimodes associated with stable periodic orbits, Comm. Math. Phys. **51** (1976), 219–242.
2. Approximate eigenfunctions of the Laplacian, J. Diff. Geom. **12** (1977), 87–100—A correction, J. Diff. Geom. **14** (1979), 487.

Randol, B.
1. Small eigenvalues of the Laplace operator on compact Riemann surfaces, Bull. Amer. Math. Soc. **80** (1974), 996–1000.
2. A Dirichlet series of eigenvalue type with applications to asymptotic estimates, Preprint CUNY 1980.
3. The Riemann hypothesis for Selberg's zeta-function and the asymptotic behavior of eigenvalues of the Laplace operator, Trans. Amer. Math. Soc. **236** (1978), 209–223.
4. On the asymptotic distribution of closed geodesics on compact Riemann surfaces, Trans. A.M.S. **233** (1977), 241–247.
5. On the analytic continuation of the Minakshisundaram-Pleijel zeta function for compact Riemann surfaces, Trans. Amer. Math. Soc. **201** (1975), 241–246.
6. The length spectrum of a Riemann surface is always of unbounded multiplicity, Preprint C.U.N.Y. New-York 1979.
7. The asymptotic behavior of a Fourier transform and the localization property of eigenfunction expansions for some partial differential operators, Trans. Amer. Math. Soc. **168** (1972), 265–271.
8. A remark on the multiplicity of the discrete spectrum of congruence groups, Proc. Amer. Math. Soc. **81** (1981), 339–340.

Rauch, J.
1. The leading wave front for hyperbolic mixed problems, Bull. Soc. Royale Sc. Liège **5–8** (1977), 156–161.
2. Perturbation theory for eigenvalues and resonances of Schrödinger hamiltonians, J. Funct. Analysis **35** (1980), 304–315.

Rauch, J.-Taylor, M.
1. Potential and scattering theory on wildly perturbed domains, J. Functional Analysis **18** (1975), 27–59.

Ray, D. B.
1. Reidemeister torsion and the Laplacian on lens spaces, Adv. in Math. **4** (1970), 101–126.

Ray, D. B.-Singer, I. M.
1. R-torsion and the Laplacian of Riemannian manifolds, Adv. in Math. **7** (1971), 145–209.
2. Analytic torsion for complex manifolds, Annals of Math. **98** (1973), 154–177.

Reed, M.-Simon, S.
1. Methods of Modern Mathematical Physics, I: Functional Analysis, II: Fourier Analysis—Self-adjoindness, III: Scattering theory, IV: Analysis of operators, Acad. Press.

Reid, W. T.
1. A comparison theorem for self-adjoint differential equation of second-order, Annals of Math. **65** (1957), 197–202.

Reilly, R. C.
1. Applications of the integral of an invariant of the Hessian, Bull. AMS **82** (1976), 579–580.
2. On the first eigenvalue of the Laplacian for compact submanifolds of Euclieam space, Commentarii Math. Helv. **52** (1977), 525–533.
3. Applications of the Hessian operator in a Riemannian manifold, Indiana U. Math. J. **26** (1977), 459–472.
4. Extrinsic estimates for $\lambda_1$, in Proc. Symp. Pure Math n° 36, AMS 1980, p. 275–278 (Geometry of the Laplace operator).
5. Geometric applications of the solvability of Neumann problems on a Riemannian manifold, Archive Rat. Mech. and Analysis **75** (1980), 23–30.
6. Mean curvature, the Laplacian and soap bubbles, Amer. Math. Monthly **89** (1982) 180–188, 197–198.

Rinke, B.-Wünsch, V.
1. Zum Huygensschen Prinzip beider skalaren Wellengleichung. Beiträge Anal. n° **18** (1981), 43–75.

Rothschild, L. L.-Wolf, J. A.
1. Eigendistribution expansions on Heisenberg groups, Indiana U. Math. J. **25** (1976), 753.

Sakai, T.
1. On eigenvalues of the Laplacian and curvature of Riemannian manifolds, Tôhoku Math. J. **23** (1971), 589–603.
2. On the spectrum of Lens spaces, Kodai Math. Sem. Rep. **27** (1975), 249–257.

Schimming, R.
1. Spectral geometry and Huygens' principle for tensor fields and differential forms, Proc. Conf. Diff. Geometry and Applic. Karlova U. 1981.

Schimming, R.-Teumer, G.
1. Spectral geometry of Laplace operators acting on tensor fields and differential forms, preprint Greifswald U. 1982.

Schmidt, W.
1. Über eine neue Methode zur Behandlung einer klasse isoperimetrischer Autgaben im Grossen, Math. Z. **47** (1942), 489–642.

Schoen, R.
1. A lower bound for the first eigenvalue of a negatively curved manifold, J. Differential Geometry, **17** (1982), 233–238.

Schoen, R.-Wolpert, S.-Yau, S. T.
1. Geometric bounds on the low eigenvalues of a compact surface, in Proc. Symp. Pure Math. **36** AMS 1980, p. 279–285 (Geometry of the Laplace operator).
2. On the first eigenvalue of a compact Riemann surface, Preprint, U.C. Berkeley.

Schwarz, A. S.
1. The partition function of degenerate quadratic functional and Ray-Singer invariants, Lett. Math. Phys. **2** (1978), 247–252.

Sealey, H. C. J.
1. On Cheng's characterization of the 2-sphere by eigenfunctions, Bull. London Math. Soc. **13** (1981), 403–404.

Seeley, R. T.

1. Complex powers of an elliptic operator, in Proc. Symp. Pure Math. n° 10, AMS 1967, p. 288–307 (Singular integrals).
2. A sharp asymptotic remainder estimate for the eigenvalue of the Laplacian in a domain of $R^3$, Adv. in Math. **29** (1978), 244–269.
3. An estimate near the boundary for the spectral function of the Laplace operator, Amer. J. Math. **102** (1980), 869–902.
4. The resolvent of an elliptic boundary problem, Amer. J. Math. **91** (1969), 889–920.
5. Analytic extension of the trace associated with elliptic boundary value problem, ibid. 963–983.
6. Norms and domains of the complex powers $A_{B^z}$, ibid. **93** (1971), 299–309.

Sekiguchi, J.
1. Eigenspaces of the Laplace-Beltrami operator on a hyperboloid, Nagoya Math. J. **79** (1980), 151–186.

Seminaire Franco-Japonais
1. Actes du Séminaire tenu à Kyoto (Oct. 81): ces actes.

Seminar on Differential Geometry.
1. Ed. by S. T. Yau, Ann. of Math. Studies **102** Princeton U. Press 1982.

Seminar on Harmonic Analysis.
1. Proceeding of the Seminar on harmonic analysis held in Pisa, April 8–17, 1980. Rend. Cir. Mat. Palermo (2) 1981. Supp. n° 1 Circolo Matematico di Palermo, Palermo 1981.

Serrin, J.
1. A symmetry problem in Potential theory, Arch. Rat. Match. Analysis **43** (1971), 304–318.

Shafii-Dehabad A.
1. Intégrales de Laplace et spectre d'une variété riemannienne sur laquelle opère un groupe d'isométries, Thèse 3ème cycle IRMA Strasbourg 1981.

Shimakura, N.
1. Stabilité locale de la première valeur propre du Laplacien pour le problème de Dirichlet, C.R.A.S. **292** (1981), 617–619.
2. La première valeur propre du Laplacien pour le problème de Dirichlet, preprint Centre de Mathématiques, Ecole Polytechnique, Palaiseau 1981.
3. La première valeur propre du Laplacien pour le problème de Dirichlet. Conférence on Partial Differential Equations. (Saint Jean de Monts, 1981) Conf. n° 14 bis **9** pp. Soc. Math. France, Paris, 1981.

Simon, U.
1. Curvature bounds for the spectrum of closed Einstein spaces, Canadian J. Math., **153** (1977), 23–27.
2. Small eigenvalues of the Laplacian, in Proc. CSSR-GDR-Polish Conference on Differential Geometry and its Applications, Sept. 80, Nove' Mesto na Moravé, CSSR.
3. Codazzi tensors and eigenfunctions, Preprint Technische U. Berlin 1981.
4. Estimates for eigenvalues of the Laplacian on compact Riemannian manifolds, preprint Technische U. Berlin 1981.
5. Estimates for eigenvalues of the Laplacian on compact Riemannian manifolds. Spectral theory of differential operators (Birmingham, Ala., 1981) pp. 371–374. North Holland Math. Studies, 55, 1981.

Simon, U.-Wissner, H.
1. Geometry of the Laplace operator, J. Univ. Kuwait Sci.
2. Geometrische Aspekte des Laplace-Operators, Jahrbuch Uberblicke Mathematik 1982, 73–92.

Singer, I. M.
1. Eigenvalues of the Laplacian and invariants of manifolds, Proc. Internat. Congress Math. 1974, p. 187–200.

Smale, S.

1. Smooth solutions of the heat and wave equations, Comment. Math. Helv. **55** (1980), 1–12.

Smith, C. L.

1. The asymptotics of the heat equation for a boundary value problem, Inventiones Math. **63** (1981), 467–494.

Smith, R. T.

1. The spherical representation of groups transitive on $S^n$, Indiana U. Math. J. **24** (1974), 307–325.

Spectral Theory of Differential Operators

1. Proceedings of the conference held at the University of Alabama, Birmingham, Ala., March 26–28, 1981. Edited by Ian W. Knowles and Roger T. Lewis. North-Holland Mathematics Studies, 55, 1981.

Sperb, R. P.

1. Untere und obere Schranke für den tiefsten Eigenwert der elastisch gestützen Membran, J. Math et de Phys. Appl. (ZAMP) **23** (1972), 231–244.

Sperner, E. Jr.

1. Spherical symmetrization and eigenvalue estimates, Math. Z. **176** (1981), 75–86.

Stochastic Differential Equations and Applications
Academic Press 1977.

Strese, H.

1. Spektren symmetrischer Räume, Math. Nachr. **98** (1980), 75–82.
2. Über den Dirac-Operator auf Grassmann Mannigfaltigkeiten, Math. Nachr. **98** (1980), 53–59.
3. Harmonic analysis of the pairs SO (n), SO(k)×SO(n-k) and SP(n), SP(k) ×SU(n-k), ibid. 61–73.
4. Spectra of symmetric spaces, ibid. 75–82.
5. Zum Spektrum des Laplace Operators auf p-Former, Math. Nachr. **106** (1982), 35–40.

Strichartz, R.

1. Analysis of the Laplacian on a complete Riemannian manifold, Preprint Cornell U., 1982.

Sulanke, S.

1. Der erste Eigenwert des Dirac-Operators auf $S^5/\Gamma$. Math. Nachr. **99** (1980) 259–271.

Sunada, T.

1. Spectrum of compact flat manifolds, Commentarii Math. Helv. **53** (1978), 613–621.
2. Trace formula for Hill's operators, Duke Math. J. **47** (1980), 529.
3. Asymptotics for path integrals defined on a Riemannian manifold: ces actes.
4. Spherical means and geodesic chains on a Riemannian manifold, Trans. A.M.S. **267** (1981), 483–501.
5. Trace formula and heat equation asymptotics for a non-positively curved manifold, Amer. J. of Math. **104** (1982), 795–812.

Suyama, Y.

1. On a problem posed by Eells-Sampson, Memoirs Fac. Sc. Kyushu U. **29** (1970), 305–315.

Svendsen, E. C.

1. The effect of submanifolds upon essential self-adjointness and deficiency indices, J. of Math. Anal. and Appl. **80** (1981), 551–565.

Swanson, C. A.

1. Asymptotic variational formulae for eigenvalues, Canadian Math. Bull. **6** (1963), 15–25.

Symes, W. W.

1. Systems of Toda type, inverse spectral problems and representation theory, Inventiones Math. **59** (1980), 13–52.

Tachibana, S.-Yamaguchi, S.
1. On the first proper space of Δ for p-forms in compact Riemannian manifolds, J. Diff. Geom. **15** (1980), 51–60.

Takahashi, T.
1. Minimal immersions of Riemannian manifolds, J. Math. Soc. Jap. **18** (1966), 380–385.

Talenti, G.
1. Elliptic equations and rearrangements, Ann. Scuola Norm. Sup. Pisa **3** (1976), 697–718.

Tamura, H.
1. Asymptotic formulas with sharp remainder estimates for bound states of Schrödinger operators I., J. Analyse Math. **40** (1981) 166–182 (1982).
2. Asymptotic formulas with sharp remainder estimates for eigenvalues of elliptic operators of second order, Duke Math. J. **49** (1982), 87–119.
3. Asymptotic formulas with sharp remainder estimates for eigenvalues of elliptic operators of second order. Proc. Japan Acad. Sc. A. Math. Sci. **57** (1981) 442–445.
4. Asymptotic formulae with sharp remainder estimates for eigenvalues of Schrödinger operators, Comm. Partial Differential Equations **7** (1982) 1–53.

Tanaka, M.
1. Compact Riemannian manifolds which are isospectral to three dimensional lens spaces, in Minimal submanifolds including geodesics, US-Japan Seminar, ed. by M. Obata, North-Holland 1979, p. 273–282.
2. Compact Riemannian manifolds which are isospectral to three dimensional lens spaces II, Proc. Fac. Sc. Tokai U. **14** (1978).

Tanaka, S.
1. Selberg's trace formula and spectrum; Osaka Math. J. **3** (1966), 205–216.

Tandai, K.-Sumitomo, T.
1. Killing vector fields on $S^n$ and the spectrum of SO(n+1)/SO(n−1)× SO(2), preprint Kyoto U.

Taniguchi, Y.
1. Normal homogeneous metric and their spectra, Osaka Math J. **18** (1981), 555–576.

Tanikawa, M.
1. The spectrum of the Laplacian of a $Z_3$-invariant domain, Proc. Japan Acad. Sc. Ser. A Math. Sc. **57** (1981), 13–18.
2. The spectrum of the Laplacian and smooth deformations of the Riemannian metric, Proc. Japan Acad. **55** (1979), 125–127.

Tanno, S.
1. Eigenvalues of the Laplacian of Riemannian manifolds, Tôhoku Math. J. **25** (1973), 391–403.
2. The spectrum of the Laplacian for 1-forms, Proc. A.M.S. **45** (1974).
3. Some differential equations on Riemannian manifolds, J. Math. Soc. Japan **30** (1978), 509–531.
4. Some metrics on a (4r+3)-sphere and spectra, Tsukuba J. Math. **4** (1980), 99–105.
5. The first eigenvalue of the Laplacian on sheres, Tôhoku Math. J. **31** (1979), 179–185.
6. A characterization of canonical sphere by the spectrum, Math. Z. 175 (1980), 267–274.
7. A characterization of a complex projective space by the spectrum, preprint Tokyo Institute of Technology.
8. Geometric expressions of eigen 1-forms of the Laplacian on spheres, (ces actes).

Taylor, M. E.
1. Fourier integral operators and harmonic analysis on compact manifolds,

in Proc. Symposia Pure Math n° 35 (Harmonic Analysis), AMS 1979.
2. Estimate on the fundamental frequency of a drum, Duke Math. J. **46** (1979), 447–453.

Toimer, G.
1. The spectrum of the Laplacian and conformally flat Riemannian manifolds. (Russian) Izv. Vyss Vcebn. Zaved Matematika (1982) 87–88.

Trudinger, N. S.
1. On the first eigenvalue of non uniformly elliptic boundary value problems, Math. Z. **174** (1980), 227–232.

Tsagas, G.
1. On the spectrum of the Bochner-Laplace operator on the 1-forms on a compact Riemannian manifold, Math. Z. **164** (1978), 153–157.
2. The spectrum of the Laplace operator for Einstein manifolds, Preprint 1980.
3. The spectrum of the Laplace operator for a special Riemannian manifold, Kodai Math. J. **4** (1981), 377–382.
4. The spectrum of the Laplace operator for a special complex manifold, in Global Differential Geometry and Analysis, Proceedings Berlin 1979, Springer L. N. in Math. n° 838, p. 233–238.

Tsagas, G.-Kochinos, K.
1. The geometry and the Laplace operator on the exterior 2-forms on a compact Riemannian manifold, Proc. Amer. Math. Soc. **73** (1979), 109–116.

Tsukada, K.
1. Locally symmetric Einstein-Kähler manifolds and spectral geometry, Tôhoku Math. J. **31** (1979), 255–259.
2. The first eigenvalue of the Laplacian on tori, Tôhoku Math. J. **33** (1981), 395–407.
3. Eigenvalues of the Laplacian on Calabi-Eckmann manifolds, J. Math. Soc. Japan **33** (1981), 673–691.
4. Hopf manifolds and spectral geometry, Trans. Amer. Math. Soc. **270** (1982), 609–621.

Tsukamoto, C.
1. Spectra of Laplace-Beltrami operators on SO(n+2) SO(2)×SO(n) and Sp(n+1)/Sp(1)×Sp(n), Osaka J. of Math. **18** (1981), 407–426.

Uchiyama, K.
1. Quelques résultats de (non) monotonie des valeurs propres du problème de Neumann, J. Faculty of Science, U. of Tokyo **24** (1977), 281–294.

Uhlenbeck, K.
1. Eigenfunctions of Laplace operators, Bull. AMS **78** (1972), 1073–1076.
2. Generic properties of eigenfunctions, Amer. J. Math. **98** (1976), 1059–1078.

Urakawa, H.
1. On the least positive eigenvalue of the Laplacian for Riemannian manifolds, Proc. Japan Acad. **53** (1977), 229–231.
2. On the least positive eigenvalue of the Laplacian for compact group manifolds, J. Math. Soc. Japan **31** (1979), 209–226.
3. On the least positive eigenvalue of the Laplacian for the compact quotient of a certain Riemannian symmetric space, Nagoya Math. J. **78** (1980), 137–152.
4. The heat equation on a compact Lie group, Osaka J. Math. **12** (1975), 285–297.
5. Analytic torsion of space forms of certain compact symmetric spaces, Nagoya Math. J. **67** (1977), 65–88.
6. How do eigenvalues of Laplacian depend upon deformations of Riemannian metrices ? (ces actes).
7. Complex Laplacians on compact complex homogeneous spaces, J. Math. Soc. Japan **33** (1981), 619–638.

8. Bounded domains which are isospectral but not isometric, Ann. Ec. Norm. Sup. **15** (1982).
9. Numerical computations of the spectra on 7-dimensional homogeneous manifolds SU(3)/T(k, 1), preprint 1982 Tohoku U.
10. Lower bounds for the eigenvalues of the fixed vibrating membrane problems, preprint Tohoku U 1982.

Vanninathan, M.
1. Homogenization of eigenvalue problems in perforated domains, Proc. Indian Acad. Sci. Math. Sci. **90** (1981) 239–271.

Varopoulos, N. Th.
1. Diffusion sur une variété riemannienne à courbure non négative, C.R. Acad. Sci. Paris Sér. I Math. **293** (1981) 213–214.
2. Fonctions harmoniques et diffusion sur une variété riemannienne, C.R. Acad. Sci. **294** (1982) 277–280.
3. Green's functions on positively curved manifolds, J. Funct. Anal. **45** (1982) 109–118.
4. The poisson kernel on positively curved manifolds, J. Funct. Anal. **44** (1981) 359–380.

Vasil'ev, D. G.
1. Asymptotics of the distribution function of the spectrum of pseudo-differential operators with parameters, Funct. Analysis and its Applications **14** (1980), 217–219.

Venkov, A. B.
1. Spectral theory of automorphic functions, the Selberg zeta function and some problems of analytic number theory and mathematical physics, Russian Math. Surveys **34** (1979), 79–153.

Véron, L.
1. Une remarque sur le spectre de $(-\Delta)^{1/2}$, preprint U. de Tours 1980.

Vignéras, M. F.
1. Variétés riemanniennes isospectrales et non isométriques, Annals of Math. (1980), 21–32.
2. Arithmétique des algèbres de quaternions, Lect. Notes in Math. n° 800, Springer 1980.
3. Quelques remarques sur la conjecture $\lambda_1 \geq \frac{1}{4}$, preprint, Université Paris 7, 1982.

Vol'pert, V. A.
1. The spectrum of an elliptic operator in an unbounded cylindrical domain. (Russian, English Summary). Dokl. Akad. Nauk. Ukrain, SSR Scr. A. 1981, 9–12, 94.

Voros, A.
1. Correspondance semi-classique et résultats exacts: cas des spectres d'opérateurs de Schrödinger homogènes, C.R. Acad. Sci. Paris, **293** (1981), 709–712.
2. Spectre de l'équation de Schrödinger et méthode BKW, Publications mathématiques d'Orsay 81, **9**. Université de Paris-Sud, Département de Mathématiques, Orsay, 1982.

Waechter, R. T.
1. On hearing the shape of a drum: an extension to higher dimensions, Proc. Camb. Phil. Soc. **72** (1972), 439–447.

Wallach, N.
1. An asymptotic formula of Gelfand and Gangolli for the spectrum of $\Gamma/G$ J. Diff. Geom. **11** (1976), 91–102.

Watanabe, T.
1. On extrapolation of minimal eigenvalue of Laplace differential equation. Bull. Aichi Uni. Ed. Natur. Sci. **30** (1981), 61–66.

Weber, H.

1. Zur Verzweigung bei einfachen Eigenwerten. Manuscripta Math. **38** (1982) 77–86.

Weinstein, A.

1. Application des opérateurs intégraux de Fourier au spectres des variétés riemanniennes, CRAS (1974).
2. Fourier integral operators, quantization, and the spectra of Riemannian manifolds, in Géométrie symplectique et physique mathématique. Colloque Int. du C.N.R.S. n° 237, Aix-en-Provence, Ed. du CNRS 1976, p. 289–298.
3. Asymptotics of eigenvalue clusters for the Laplacian plus a potential, Duke Math. J. **44** (1977), 883–892.
4. Eigenvalues of a Laplacian plus a potential, Proc. I.C.M. Helsinki **2** (1978), 803–805.
5. On the $L^4$-norm of spherical harmonics, Proc. Symp. in Pure Math. n° 36, Amer. Math. Soc. Providence 1980 (Geometry of Laplace Operator).

Weyl, H.

1. Über die asymptotische Verteilung der Eigenwerte, Nachr. der Königl. Ges. d. Wiss. zu Göttingen, (1911), 110–117.

Widom, H.

1. Asymptotics of compressions to spectral subspaces of the Laplacian, in Proc. Symp. Pure Math. n° 36, AMS 1980, p. 319–323 (Geometry of the Laplace operator).
2. The Laplace operator with potential on the two-sphere, Adv. in Math. **31** (1979), 63–66.
3. Eigenvalue distribution theorem in certain homogeneous spaces, J. Funct. Analysis **32** (1979).

Wodzicki, M.

1. Spectral asymmetry and zeta functions. Invent. Math. **66** (1982) 115–135.

Wolpert, S.

1. The eigenvalue spectrum as moduli for compact Riemann surfaces, Bull. AMS **83** (1977), 1306–1308.
2. The eigenvalue spectrum as moduli for flat tori, Trans. AMS, **244** (1978), 313–321.
3. The length spectra as moduli for compact Riemann surfaces, Annals of Math. **109** (1979), 323–351.
4. On the variational theory of the Laplacian for hyperbolic surfaces, preprint U. of Maryland 1979.
5. Eigenvalues of the Maass wave forms, U. of Maryland Preprint 1978.

Xavier, F.

1. Spectral theory of the Laplace-Beltrami operator on complete non-compact manifolds, Thesis U. of Rochester 1977.

Yamaguchi, S.

1. Spectra of flag manifolds, Memoirs Fac. Sc. Kyushu U. **33** (1979), 95–112.
2. Some remarks on the eigenvalues of the Laplace operators of certain compact simply connected Riemannian manifolds. preprint Yamaguchi U., 1982.

Yang, P. C.-Yau, S. T.

1. Eigenvalues of the Laplacian of compact Riemann surfaces and minimal submanifolds, Annali della Scuola Norm. Sup. di Pisa **7** (1980), 55–63.

Yau, S. T.

1. Isoperimetric constants and the first eigenvalue of a compact Riemannian manifold, Ann. Scien. Ecole Norm. Sup. **8** (1975), 487–507.
2. Some function theoretic properties of complete Riemannian manifolds and their applications in geometry, Indiana U. Math. J. **25** (1976), 659–670.
3. Survey on partial differential equations in differential geometry, Seminar on Differential Geometry, Ed. by S. T. Yau, Annals of Math Studies n° 102, Princeton U. Press 1982.
4. Problem section, ibid.

Zalcman, L.
1. Offbeat integral geometry, Amer. Math. Monthly **87** (1980), 161–175.
Zograf, P. G.
1. Fuchsian groups and small eigenvalues of the Laplace operator. Russian studies in Topology. IV. Zap Naučn. Sem. Leningrad. Otdel. Mat. Inst. Steklov (LOMI) **122** (1982) 24–29, 163.
Zucker, S.
1. Estimates for the classical parametrix for the Laplacian, Manuscripta Math. **24** (1978), 9–29.

Université de Savoie
Service de Mathématiques
BP 1104
73011 Chambery Cedex
France

Université Paris 7
U.E.R. de Mathématiques
L.A. au C.N.R.S. n° 212
75251 Paris Cedex 05
France

# APPENDIX C

## A COMPLEMENT TO APPENDIX B

# APPENDIX C

## Complement to Appendix B

The following (obviously not exhaustive) list of references is an addendum to the bibliography [B-B] given in Appendix B. We do not intend to give an up to date 1986 - version of [B-B], but we think that it might be useful to point out some contributions to spectral geometry which appeared in the last four years. Some papers are major contributions; some others are less important, but give interesting developments to the subject.

The numbers after each paper refer to the classification given in [B-B].


ALVAREZ-GAUMÉ, L. - Supersymmetry and the Atiyah-Singer index theorem, Comm. Math. Phys. 90 (1983), 161-173 [§4.1, 5.1].

ANDERSON, M.T. - The Dirichlet problem at infinity for manifolds of negative curvature, J. Diff. Geom. 18 (1983), 701-721 [§10].

ANNÉ, C. - Spectre du laplacien et limites de variétés avec perte de dimension, Preprint Institut Fourier, Grenoble nº 45, 1985 [§7.1].

ATIYAH, M.F. - DONNELLY, H. - SINGER, I.M. - Eta invariants, signature defects of cusps, and values of L-functions, Ann. of Math. 118 (1983), 131-177; Addendum, ibid. 119 (1984), 635-637 [§5.3].

BARBASCH, D. - MOSCOVICI, H. - $L^2$-index and the Selberg trace formula, J. Funct. Anal. 53 (1983), 151-201 [§9.1].

BÉRARD, P. - From vanishing theorems to estimating theorems: the Bochner technique revisited, Preprint Univ. de Savoie 1986 [§11.3].

BÉRARD, P. - GALLOT, S. - Remarques sur quelques estimées géométriques explicites, C.R. Acad. Sci. 297 (1983), 185-188 [§11.3].

BERLINE, N. - VERGNE, M. - A computation of the equivariant index of the Dirac operator, Bull. Soc. Math. France to appear [§4.1, 5.1].

BESSON, G. - Comportement asymptotique des valeurs propres du laplacien dans un domaine avec un trou, Bull. Soc. Math. France 113 (1985), 211-230 [§7.1].

BISMUT, J.M. - The Atiyah-Singer theorems: a probabilistic approach I. The index theorem, J. Funct. Anal. 57 (1984), 56-99; II. The Lefschetz fixed point formulas, ibid. 329-348 [§4.1, 5.1, 12].

BISMUT, J.M. - The Atiyah-Singer index theorems for families of Dirac operators: two heat equation proofs, Preprint Univ. Paris Sud-Orsay, 1985, [§4.1, 5.1, 12]

BROOKS, R. - On the spectrum of non compact manifolds with finite volume, Math. Z. 187 (1984), 425-432 [§10].

BROWDER, F.E. - Coïncidence theorems, minimax theorems, and variational inequalities, in Conference in Modern Analysis and Probability (New Haven, Conn 1982), Contemp. Math. nº 26, Amer. Math. Soc. 1984, p. 67-80 [§1, 2.2]

BRÜNING, J. - On the compactness of isospectral potentials, Comm. Partial Diff. Eq. 9 (1984), 687-698 [§ 6].

BRÜNING, J. - HEINTZE, E. - Spektrale Starrheit gewisser Drehflächen, Math. Ann. 269 (1984), 95-101 [§ 6].

BUSER, P. - Isospectral Riemann surfaces, Ann. Inst. Fourier, to appear [§ 6].

CHEEGER, J. - Spectral geometry of singular Riemannian spaces, J. Diff. Geom. 18 (1983), 575-657 [§ 4,5].

CAFFARELLI, L. - SPRUCK, J. - Convexity properties of solutions to some classical variational problems, Comm. Partial Diff. Eq. 7 (1982), 1337-1379 [§ 12].

CHOI, H.I. - WANG, A.N. - A first eigenvalue estimate for minimal hypersurfaces, J. Diff. Geom. 18 (1983), 551-568 [§ 11.1].

COLIN DE VERDIÈRE, Y. - Pseudo laplaciens, I. Ann. Inst. Fourier 32 (1982), 275-286; II. ibid 33 (1983), 87-113 [§ 1].

. Sur la multiplicité de la première valeur propre non nulle du laplacien, Preprint Institut Fourier, Grenoble 1985 [§ 11.1, 12].

. L'asymptotique de Weyl pour les bouteilles magnétiques, Preprint Institut Fourier, Grenoble 1985 [§ 4.3, 5.5].

COURTOIS, G. - Estimations du noyau de l'opérateur de la chaleur et du noyau de Green d'une variété riemannienne. Applications aux variétés privées d'un $\epsilon$-tube, C.R. Acad. Sci. 1986 to appear [§7.1].

DETURCK, D. - GORDON, C. - Isospectral deformations I: Riemannian Structures on two steps nilspaces, to appear [§ 6].

DODZIUK, J. - Difference equations, isoperimetric inequality and transience of certain random walks, Trans. Amer. Math. Soc. 284 (1984), 787-794 [§2.2, 11.3, 12].

. Every covering of a compact Riemann surface of genus greater than one carries a non trivial $L^2$-harmonic differential, Acta Math. 152 (1984),

49-56 [§ 10].

DONNELLY, H. - LI, P. - Heat equation and compactifications of complete Riemannian manifolds, Duke Math. J. 51 (1984), 667-673 [§10].

EICHHORN, J. - The influence of bounded and unbounded geometry upon the spectrum, in Diff. Geom. Ed by Kowalski, Prague 1984, p. 21-32 [§ 5,12].

EL SOUFI, A. - ILIAS, S. - Le volume conforme et ses applications d'après Li et Yau, in Séminaire "Théorie Spectrale et Géométrie", 1984-1985, Institut Fourier, Grenoble [§ 11, 12].

GALLOT, S. - Inégalités isopérimétriques, courbure de Ricci et invariants géométriques I., C.R. Acad. Sci. 296 (1983), 333-337; II., ibid. 365-369 [§11.3, 12].

GETZLER, E. - Pseudo differential operators on supermanifolds and the Atiyah-Singer index theorem, Comm. Math. Phys. 92 (1983), 163-178 [§ 4.1, 5.1].

GILKEY, P.B. - Invariance theory, the heat equation and the Atiyah-Singer index theorem, Lect. Notes 11, Publish or Perish Inc. 1984 [§ 4.1, 5.1].

GORDON, C.S. - WILSON, E.N. - Isospectral deformations of compact solvmanifolds, J. Diff. Geom. 19 (1984), 241-256 [§ 6].

GROMOV, M. - LAWSON, H.B. - Positive scalar curvature and the Dirac operator on complete Riemannian manifolds, Inst. Hautes Etudes Sci. Publ. Math. 58 (1983), 83-196 [§ 4, 5, 12].

GUILLEMIN, V. - URIBE, A. - Spectral properties of a certain class of complex potentials, Trans. Amer. Math. Soc. 279 (1983), 759-771 [§ 6, 5.5].

GUREJEV, T.E. - SAFAROV, Yu.G. - Precise asymptotics of the spectrum for the Laplace operator on manifolds with periodic geodesics, LOMI Preprints E-1-86, USSR Academy of Sciences, Steklov Math. Inst. Leningrad [§ 4.3, 5].

HEJHAL, D.A. - The Selberg trace formula for PSL(2, ℝ), Vol. 2, Lect. Notes in Math. 1001, Springer 1983 [§ 9].

HENNIART, G. - Les inégalités de Morse d'après E. Witten (J. Diff. Geom. 17 (1982), 661-692), in Séminaire Bourbaki, Exp. 617 1983-1984 [§ 2.2, 5, 12].

HIJAZI, O. - Opérateurs de Dirac sur les variétés riemanniennes: minoration des valeurs propres, Thèse de 3ieme cycle, Université Paris 6, 1984 [§ 11].

HINZ, A. - Obere Schranken für Eigenfunktionen eines Operators $-\Delta+q$, Math. Z. 185 (1984), 291-304 [§ 5.5, 11].

IVRII, V. - Precise spectral asymptotics for elliptic operators acting in fiberings over manifolds with boundary, Lect. Notes in Math. 1100, Springer 1984 [§ 4.3].

JENNI, F. - Über den ersten Eigenwert des Laplace-Operators auf ausgewählten Beispielen kompakter Riemannscher Flächen, Comment. Math. Helv. 59 (1984), 193-203 [§ 3, 11].

KASUE, A. - Applications of Laplacian and Hessian comparison theorems, in Geometry of geodesics and related topics, p. 333-386, Adv. Studies Pure Math. 3, North-Holland 1984 [§ 11].

KAWOHL, B. - Rearrangements and convexity of level sets in partial differential equations, Lect. Notes in Math. nº 1150, Springer 1985 [§ 11.3, 12].

KUTTLER, J.R. - SIGILLITO, V.G. - Eigenvalues of the Laplacian in two dimensions, Siam Review 26 (1984), 163-193 [§ 4, 11, Review].

LI, P. - YAU, S.T. - On the Schrödinger equation and the eigenvalue problem, Comm. Math. Phys. 88 (1983), 309-318 [§ 4,5,10].

MARZIVI, S. - MELROSE, R. - Spectral invariants of convex planar regions, J. Diff. Geom. 17 (1982), 475-502 [§ 5.1, 6].

MULLER, W. - Spectral theory for Riemannian manifolds with cusps and a related trace formula, Math. Nachr. 111 (1983), 197-288 [§ 11, § 4,5].

PETKOV,V.M. - Propriétés génériques des rayons réfléchissants et applications aux problèmes spectraux, Séminaire Bony-Sjöstrand-Meyer 1984-1985, Exp. XII, Ecole Polytechnique Palaiseau 1985 [§ 5].

REMPEL, S. - SCHMITT, T. - Pseudo differential operators and the index theorem on supermanifolds, Math. Nachr. 111 (1983), 153-175 [§ 4,5].

SCHOEN, R. - A lower bound for the first eigenvalue of a negatively curved manifold, J. Diff. Geom. 17 (1982), 233-238 [§ 11.1].

STANTON, N.K. - The heat equation in several complex variables, Bull. Amer. Math. Soc. 11 (1984), 65-84 [§ 2.2, 4].

STRICHARTZ, R.S. - Mean value properties of the Laplacian via spectral theory, Trans. Amer. Math. Soc. 284 (1984), 219-228 [§ 2].

SULLIVAN, D. - The Dirichlet problem at infinity for a negatively curved manifold, J. Diff. Geom. 18 (1983), 723-732 [§ 10].

SUNADA, T. - Riemannian coverings and isospectral manifolds, Annals of Math. 121 (1985), 169-186 [§ 6]

TAMURA, H. - Asymptotic formulas with sharp remainder estimates for bound states of Schrödinger operators, J. Analyse Math. 41 (1982), 85-108 [§ 4.3, 5.5].

TELEMAN, N. - The index of signature operators on Lipschitz manifolds, Inst. Hautes Etudes. Sci. Publ. Math. 58 (1983), 39-78 [§ 4,5].

VAINBERG, B.R. - Complete asymptotic expansion of the spectral function of second-order elliptic operators in $\mathbb{R}^n$ (Russian) Math. Sb. 123 (165) (1984), 195-211 [§ 4.3].

VAROPOULOS, N. - Chaines de Markov et inégalités isopérimétriques, C.R. Acad. Sci. 298 (1984), 233-236 [§ 11.3, 12].

WODZICKI, M. - Local invariants of spectral asymmetry, Invent. Math. 75 (1984), 143-177 [§ 5].

ZELDITCH, S. - Eigenfunctions on compact Riemannian surfaces of genus $g \geq 2$, Preprint 1984 [§ 12].

## REFERENCES


[AN] AGMON, S. - Lectures on Elliptic boundary value problems, Van Nostrand 1965.

[AZ] ARONSZAJN, N. - A unique continuation theorem for solutions of elliptic partial differential equations or inequalities of second order, J. Math. Pures Appl. 36 (1957), 235-249.

[B-B] BERARD, P. - BERGER, M. - Le Spectre d'une variété riemannienne en 1982, in Spectra of Riemannian Manifolds, Kaigai Publications 1983, p. 139-194 (see Appendix B).

[B-B-G1] BERARD, P. - BESSON, G. - GALLOT, S. - Sur une inégalité isopérimétrique qui généralise celle de Paul Lévy-Gromov, Inventiones Math. 80 (1985), 295-

[B-B-G2] BERARD, P. - BESSON, G. - GALLOT, S. - Sur la fonction isopérimétrique des variétés riemanniennes compactes sans bord, In preparation 1985.

[B-B-G3] BERARD, P. - BESSON, G. - GALLOT, S. - In preparation 1985.

[B-C] BISHOP, R.L. - CRITTENDEN, R.J. - Geometry of Manifolds, Acad. Press. 1964.

[BD] BERARD, P. - Heat and wave operators on compact Riemannian manifolds, Notas de Curso nº 13, Universidade Federal de Pernambuco, Recife (Brazil) 1978.

[BD1] BERARD, P. - From vanishing theorems to estimating theorems: the Bochner technique revisited, Preprint 1985.

[BD2] BERARD, P. - A propos de "Eigenvalues in Riemannian Geometry" de I. Chavel, Gazette des Mathématiciens, Soc. Math. France 1986 to appear.

[BE] BANDLE, C. - Isoperimetric inequalities and applications, Pitman 1980.

[B-G] BERARD, P. GALLOT, S. - Inégalités isopérimétriques pour l'équation de la chaleur et application à l'estimation de quelques invariants, Exposé nº XV, Séminaire Goulaouic-Meyer-Schwartz 1983-1984, Ecole Polytechnique Palaiseau 1984.

Remarques sur quelques estimées géométriques explicites, C.R. Acad. Sci. Paris 297 (1983), 185-188.

[B-G-M] BERGER, M. - GAUDUCHON, P. - MAZET, E. - Le spectre d'une variété riemannienne, Lecture Notes in Mathematics nº 194, Springer 1971.

[B-J-S] BERS, L. - JOHN, F. - SCHECHTER, M. - Partial differential equations, Interscience 1964.

[B-M] BERARD, P. - MEYER, D. - Inégalités isopérimétriques et applications, Ann. Sci. Ec. Norm. Sup., Paris 15 (1982), 513-542.

[BN] BERENSTEIN, C. - Autovalores del Laplaciano y geometria, VII ELAM, Fondo Editorial, acta cientifica venezolana, Caracas 1984.

[BR] BUSER, P. - On Cheeger's inequality $\lambda_1 \geq h^2/4$, Proceedings of Symposia in Pure Math. nº 36 (The geometry of the Laplace operator) Amer. Math. Soc. 1980, p. 29-77.

[BS] BESSE, A.L. - Manifolds all of whose geodesics are closed, Springer 1978.

[BZ] BREZIS, H. - Analyse fonctionnelle: Théorie et Applications, Masson 1983.

[C-E] CHEEGER, J. - EBIN, D. - Comparison theorems in Riemannian geometry, North-Holland 1975.

[CG] CHENG, S.Y. - Eigenvalue comparison theorems and its geometric applications, Math. Z. 143 (1975), 289-297.

[C-H] COURANT, R. - HILBERT, D. - Methods of mathematical physics, Vol I, J. Wiley & Sons 1953.

[CL] CHAVEL, I. - Eigenvalues in Riemannian Geometry, Academic Press 1984.

[CO] CARMO, M. do - Geometria Riemaniana - Escola de Geometria Diferencial Universidade Federal do Ceará Julho 1978.

[DK] DODZIUK, J. - Eigenvalues of the Laplacian and the heat equation, Amer. Math. Monthly 88 (1981), 686-695.

[D-M] DYM, H. - Mc KEAN, H.P. - Fourier series and integrals, Académic Press 1972.

[E-L] EELLS, J. and LEMAIRE, L. - Selected topics in harmonic maps, C.B.M.S. Regional Conf. Series 50, Amer. Math. Soc. 1983.

[E S] EELLS, J. - Elliptic operators on manifolds, Proceedings Summer Course Complex Analysis I.C.T.P. Trieste 1975 International Atomic Energy Agency 1976 Vol. 1, p. 95-152.

[FO] FIGUEIREDO, D.G. - Análise de Fourier e equações diferenciais parciais, Projeto Euclides, I.M.P.A. CNPq, Editora E. Blücher, Ltda., 1977.

[FR] FEDERER, H. - Geometric measure theory, Springer 1969.

[GA1] GALLOT, S. - A Sobolev inequality and some geometric applications, in Spectra of Riemannian manifolds, Kaigai Publ. 1983, p. 45-55.

[GA2] GALLOT, S. - Inégalités isopérimétriques, courbure de Ricci et invariants géométriques I and II, C.R. Acad. Sci. Paris <u>296</u> (1983), 333-336 and 365-368.

[GL1] GUILLEMIN, V. - Lectures on spectral theory of elliptic operators, Duke Math. J. <u>44</u> (1977), 485-517.

[GL2] GUILLEMIN, V. - Some classical theorems in spectral theory revisited, in Seminar on singularities of solutions of partial differential equations, ed. by L. Hörmander, Annals of Math. Studies nº 91, Princeton Univ. Press 1979.

[G-M] GALLOT, S. - MEYER, D. - Opérateur de courbure et Laplacien des formes différentielles d'une variété riemannienne, J. Math. Pures et Appl. <u>54</u> (1975), 259-284.

[GN] GARABEDIAN, P.R. - Partial differential equations, J. Wiley and Sons, 1964.

[G-S] GUILLEMIN, V. - STERNBERG, S. - Geometric asymptotics, Amer. Math. Soc. 1977.

[G-T] GILBARG, D. - TRUDINGER, N.S. - Elliptic partial differential equations of second order, Springer 1977.

[GV1] GROMOV, M. - Paul Levy's isoperimetric inequality, Pretirage IHES 1980.

[GV2] GROMOV, M. - Structures métriques pour les variétés riemanniennes, rédigé par J. Lafontaine et P. Pansu, Cedic-Nathan 1980.

[GY] GILKEY, P.B. - The index theorem and the heat equation, Publish or Perish Inc, 1974.

[HF] HOPF, H. - Differential geometry in the large, Lec. Notes in Math. nº 1000, Springer 1983.

[H-S-U] HESS, H. - SCHRADER, R. - UHLENBROCK, D.A. - Kato's inequality and the spectral distribution of Laplacians on compact Riemannian manifolds, J. Diff. Geom. 15 (1980), 27-38.

[KG] KLINGENBERG, W. - Riemannian geometry, de Gruyter Studies in Math. nº 1, de Gruyter 1982.

[KL] KAWOHL, B. - On rearrangements, symmetrization and maximum principles, Preprint Institut für Angewandte Mathematik, Universität Erlangen D 8520 ERLANGEN, L.N. 1150 Springer.

[KL] KAWOHL, B. - Rearrangements and Convexity of Level Sets in PDE, Lecture Notes in Mathematics 1150 Springer-Verlag, 1985.

[K-N] KOBAYASHI, S. - NOMIZU, K. - Foundations of differential geometry, Vol. I and II, Interscience Publ. 1963.

[KO] KATO, T. - Perturbation theory for linear operators, Springer 1965.

[K-S] KUTTLER, J.R. - SIGILLITO, V.G. - Eigenvalues of the Laplacian in two dimensions, SIAM Review, 26 (1984) 163-193.

[LG] LANG, S. - Differential manifolds, Addison-Wesley 1972.

[LZ] LICHNEROWICZ, A. - Géométrie des groupes de transformations, Dunod 1958.

[ME1] MEYER, D. - Minoration de la première valeur propre non nulle du problème de Neumann sur les variétés riemanniennes à bord, Preprint Inst. Fourier Grenoble nº 23, 1985.

[ME2] MEYER, D. - Une inégalité de géométrie hilbertienne et ses applications à la géométrie riemannienne, Preprint 1984.

[MO] MOSSINO, J. - Inégalités isopérimétriques et applications en physique, Hermann 1984.

[MR] MILNOR, J. - Morse theory, Annals of Math. Studies nº 51, Princeton Univ. Press 1963.

[M-S] MERCURI, A. - RIGAS, A. Curvatura e topologia, Escola de Geometria Diferencial Univ. Federal do Ceará, Julho 1978.

[M-T] MOSSINO, J. - RAKOTOSON, J.M. - Isoperimetric inequalities in parabolic equations, Ann. Scuola Normale Sup. Pisa, to appear.

[NN] NARASIMHAN, R. - Analysis on real and complex manifolds, Masson-North-Holland 1968.

[ON] OSSERMAN, R. - The isoperimetric inequality, Bull. Amer. Math. Soc. 84 (1978), 1182-1238.

[PE] PAYNE, L.E. - Isoperimetric inequalities and their applications, SIAM Review 9 (1967), 453-488.

[P-S] POLYA, G. - SZEGO, G. - Isoperimetric inequalities in mathematical physics, Annals of Math. Studies 27, Princeton University Press 1951.

[P-W] PROTTER, M.H. - WEINBERGER, H.F. - Maximum principles in differential equations, Prentice-Hall 1967.

[PY] PETROVSKY, I.G. - Lectures on partial differential equations, Interscience Publ. 1954.

[RH] RAYLEIGH, J.W. STRUTT (Lord) - The theory of sound, Dover 1945.

[RM] de RHAM, G. - Géométrie differentiable, Hermann 1960.

[R-S] REED, M. - SIMON, B. - Methods of Mathematical Physics, Vol. I - IV, Academic Press 1975.

[SI] SAKAI, T. - Comparison and finiteness theorems in Riemannian geometry, Advanced studies in Pure Math. 3, 1984, Geometry of Geodesics and Related Topics p. 125-181. North Holland - Kinokuniya 1984.

[SK] SPIVAK, M. - A comprehensive introduction to differential geometry, Publish or Perish Inc. (Vol. I to V).

[SR] SOTOMAYOR, J. - Lições de equações diferenciais ordinárias, Projeto Euclides, I.M.P.A., CNPq, Livros Técnicos e Científicos Editora S.A. 1979.

[SW] SHOWALTER, R.E. - Hilbert space methods for Partial Differential Equations, Pitman 1977.

[TI] TALENTI, G. - Elliptic equations and rearrangements, Ann. Scuola Norm. Sup. Pisa <u>3</u> (1976), 697-718.

[TL] TRUESDELL, C. - The influence of elasticity on analysis: the classic heritage. Bull. Amer. Math. Soc. <u>9</u> (1983), 293-310.

[TR1] THAYER, F.J. - Théorie spectrale, Monografias de Matemática nº 36, IMPA 1982.

[TR2] THAYER, J. - Notes on partial differential equations, Monografias de Matemática nº 34, IMPA 1980.

[TS] TREVES, F. - Basic linear partial differential equations, Academic Press 1975.

[TSG] THEORIE SPECTRALE ET GEOMETRIE, Séminaire 1983-1984, Exposé nº VIII, Théorèmes de finitude en géométrie riemannienne par P. Bérard et G. Besson, Universités Chambéry-Grenoble 1984.

[UA] URAKAWA, H. - Stability of harmonic maps and eigenvalues of Laplacian, Preprint Universität Bonn (F.R.G.), 1984.

[WA] WARNER, F. - Foundations of differentiable manifolds and Lie groups, Scott, Foresman and Co, 1971.

[WR] WEINBERGER, H.F. - A first course in partial differential equations, Blaisdell, Waltham Mass. 1965.

[WS] WELLS, R. - Differential analysis on complex manifolds, Graduate Texts in Math. 65 Springer (1980)

[YU] YAU, S.T. - Seminar on differential geometry, ed. by S.T. YAU, Annals of Math. Studies nº 102, Princeton Univ. Press 1982.

## SUBJECT INDEX

(numbers refer to pages)

INDEX OF NOTATIONS

(numbers refer to pages)

- * * * -

Vol. 1062: J. Jost, Harmonic Maps Between Surfaces. X, 133 pages. 1984.

Vol. 1063: Orienting Polymers. Proceedings, 1983. Edited by J.L. Ericksen. VII, 166 pages. 1984.

Vol. 1064: Probability Measures on Groups VII. Proceedings, 1983. Edited by H. Heyer. X, 588 pages. 1984.

Vol. 1065: A. Cuyt, Padé Approximants for Operators: Theory and Applications. IX, 138 pages. 1984.

Vol. 1066: Numerical Analysis. Proceedings, 1983. Edited by D.F. Griffiths. XI, 275 pages. 1984.

Vol. 1067: Yasuo Okuyama, Absolute Summability of Fourier Series and Orthogonal Series. VI, 118 pages. 1984.

Vol. 1068: Number Theory, Noordwijkerhout 1983. Proceedings. Edited by H. Jager. V, 296 pages. 1984.

Vol. 1069: M. Kreck, Bordism of Diffeomorphisms and Related Topics. III, 144 pages. 1984.

Vol. 1070: Interpolation Spaces and Allied Topics in Analysis. Proceedings, 1983. Edited by M. Cwikel and J. Peetre. III, 239 pages. 1984.

Vol. 1071: Padé Approximation and its Applications, Bad Honnef 1983. Prodeedings. Edited by H. Werner and H.J. Bünger. VI, 264 pages. 1984.

Vol. 1072: F. Rothe, Global Solutions of Reaction-Diffusion Systems. V, 216 pages. 1984.

Vol. 1073: Graph Theory, Singapore 1983. Proceedings. Edited by K.M. Koh and H.P. Yap. XIII, 335 pages. 1984.

Vol. 1074: E.W. Stredulinsky, Weighted Inequalities and Degenerate Elliptic Partial Differential Equations. III, 143 pages. 1984.

Vol. 1075: H. Majima, Asymptotic Analysis for Integrable Connections with Irregular Singular Points. IX, 159 pages. 1984.

Vol. 1076: Infinite-Dimensional Systems. Proceedings, 1983. Edited by F. Kappel and W. Schappacher. VII, 278 pages. 1984.

Vol. 1077: Lie Group Representations III. Proceedings, 1982–1983. Edited by R. Herb, R. Johnson, R. Lipsman, J. Rosenberg. XI, 454 pages. 1984.

Vol. 1078: A.J.E.M. Janssen, P. van der Steen, Integration Theory. V, 224 pages. 1984.

Vol. 1079: W. Ruppert. Compact Semitopological Semigroups: An Intrinsic Theory. V, 260 pages. 1984

Vol. 1080: Probability Theory on Vector Spaces III. Proceedings, 1983. Edited by D. Szynal and A. Weron. V, 373 pages. 1984.

Vol. 1081: D. Benson, Modular Representation Theory: New Trends and Methods. XI, 231 pages. 1984.

Vol. 1082: C.-G. Schmidt, Arithmetik Abelscher Varietäten mit komplexer Multiplikation. X, 96 Seiten. 1984.

Vol. 1083: D. Bump, Automorphic Forms on GL (3,IR). XI, 184 pages. 1984.

Vol. 1084: D. Kletzing, Structure and Representations of Q-Groups. VI, 290 pages. 1984.

Vol. 1085: G.K. Immink, Asymptotics of Analytic Difference Equations. V, 134 pages. 1984.

Vol. 1086: Sensitivity of Functionals with Applications to Engineering Sciences. Proceedings, 1983. Edited by V. Komkov. V, 130 pages. 1984

Vol. 1087: W. Narkiewicz, Uniform Distribution of Sequences of Integers in Residue Classes. VIII, 125 pages. 1984.

Vol. 1088: A.V. Kakosyan, L.B. Klebanov, J.A. Melamed, Characterization of Distributions by the Method of Intensively Monotone Operators. X, 175 pages. 1984.

Vol. 1089: Measure Theory, Oberwolfach 1983. Proceedings. Edited by D. Kölzow and D. Maharam-Stone. XIII, 327 pages. 1984.

Vol. 1090: Differential Geometry of Submanifolds. Proceedings, 1984. Edited by K. Kenmotsu. VI, 132 pages. 1984.

Vol. 1091: Multifunctions and Integrands. Proceedings, 1983. Edited by G. Salinetti. V, 234 pages. 1984.

Vol. 1092: Complete Intersections. Seminar, 1983. Edited by S. Greco and R. Strano. VII, 299 pages. 1984.

Vol. 1093: A. Prestel, Lectures on Formally Real Fields. XI, 125 pages. 1984.

Vol. 1094: Analyse Complexe. Proceedings, 1983. Edité par E. Amar, R. Gay et Nguyen Thanh Van. IX, 184 pages. 1984.

Vol. 1095: Stochastic Analysis and Applications. Proceedings, 1983. Edited by A. Truman and D. Williams. V, 199 pages. 1984.

Vol. 1096: Théorie du Potentiel. Proceedings, 1983. Edité par G. Mokobodzki et D. Pinchon. IX, 601 pages. 1984.

Vol. 1097: R.M. Dudley, H. Kunita, F. Ledrappier, École d'Éte de Probabilités de Saint-Flour XII – 1982. Edité par P.L. Hennequin. X, 396 pages. 1984.

Vol. 1098: Groups – Korea 1983. Proceedings. Edited by A.C. Kim and B.H. Neumann. VII, 183 pages. 1984.

Vol. 1099: C.M. Ringel, Tame Algebras and Integral Quadratic Forms. XIII, 376 pages. 1984.

Vol. 1100: V. Ivrii, Precise Spectral Asymptotics for Elliptic Operators Acting in Fiberings over Manifolds with Boundary. V, 237 pages. 1984.

Vol. 1101: V. Cossart, J. Giraud, U. Orbanz, Resolution of Surface Singularities. Seminar. VII, 132 pages. 1984.

Vol. 1102: A. Verona, Stratified Mappings – Structure and Triangulability. IX, 160 pages. 1984.

Vol. 1103: Models and Sets. Proceedings, Logic Colloquium, 1983, Part I. Edited by G.H. Müller and M.M. Richter. VIII, 484 pages. 1984.

Vol. 1104: Computation and Proof Theory. Proceedings, Logic Colloquium, 1983, Part II. Edited by M.M. Richter, E. Börger, W. Oberschelp, B. Schinzel and W. Thomas. VIII, 475 pages. 1984.

Vol. 1105: Rational Approximation and Interpolation. Proceedings, 1983. Edited by P.R. Graves-Morris, E.B. Saff and R.S. Varga. XII, 528 pages. 1984.

Vol. 1106: C.T. Chong, Techniques of Admissible Recursion Theory. IX, 214 pages. 1984.

Vol. 1107: Nonlinear Analysis and Optimization. Proceedings, 1982. Edited by C. Vinti. V, 224 pages. 1984.

Vol. 1108: Global Analysis – Studies and Applications I. Edited by Yu.G. Borisovich and Yu.E. Gliklikh. V, 301 pages. 1984.

Vol. 1109: Stochastic Aspects of Classical and Quantum Systems. Proceedings, 1983. Edited by S. Albeverio, P. Combe and M. Sirugue-Collin. IX, 227 pages. 1985.

Vol. 1110: R. Jajte, Strong Limit Theorems in Non-Commutative Probability. VI, 152 pages. 1985.

Vol. 1111: Arbeitstagung Bonn 1984. Proceedings. Edited by F. Hirzebruch, J. Schwermer and S. Suter. V, 481 pages. 1985.

Vol. 1112: Products of Conjugacy Classes in Groups. Edited by Z. Arad and M. Herzog. V, 244 pages. 1985.

Vol. 1113: P. Antosik, C. Swartz, Matrix Methods in Analysis. IV, 114 pages. 1985.

Vol. 1114: Zahlentheoretische Analysis. Seminar. Herausgegeben von E. Hlawka. V, 157 Seiten. 1985.

Vol. 1115: J. Moulin Ollagnier, Ergodic Theory and Statistical Mechanics. VI, 147 pages. 1985.

Vol. 1116: S. Stolz, Hochzusammenhängende Mannigfaltigkeiten und ihre Ränder. XXIII, 134 Seiten. 1985.

Vol. 1117: D.J. Aldous, J.A. Ibragimov, J. Jacod, Ecole d'Été de Probabilités de Saint-Flour XIII – 1983. Édité par P.L. Hennequin. IX, 409 pages. 1985.

Vol. 1118: Grossissements de filtrations: exemples et applications. Seminaire, 1982/83. Edité par Th. Jeulin et M. Yor. V, 315 pages. 1985.

Vol. 1119: Recent Mathematical Methods in Dynamic Programming. Proceedings, 1984. Edited by I. Capuzzo Dolcetta, W.H. Fleming and T. Zolezzi. VI, 202 pages. 1985.

Vol. 1120: K. Jarosz, Perturbations of Banach Algebras. V, 118 pages. 1985.

Vol. 1121: Singularities and Constructive Methods for Their Treatment. Proceedings, 1983. Edited by P. Grisvard, W. Wendland and J.R. Whiteman. IX, 346 pages. 1985.

Vol. 1122: Number Theory. Proceedings, 1984. Edited by K. Alladi. VII, 217 pages. 1985.

Vol. 1123: Séminaire de Probabilités XIX 1983/84. Proceedings. Edité par J. Azéma et M. Yor. IV, 504 pages. 1985.

Vol. 1124: Algebraic Geometry, Sitges (Barcelona) 1983. Proceedings. Edited by E. Casas-Alvero, G.E. Welters and S. Xambó-Descamps. XI, 416 pages. 1985.

Vol. 1125: Dynamical Systems and Bifurcations. Proceedings, 1984. Edited by B.L.J. Braaksma, H.W. Broer and F. Takens. V, 129 pages. 1985.

Vol. 1126: Algebraic and Geometric Topology. Proceedings, 1983. Edited by A. Ranicki, N. Levitt and F. Quinn. V, 523 pages. 1985.

Vol. 1127: Numerical Methods in Fluid Dynamics. Seminar. Edited by F. Brezzi, VII, 333 pages. 1985.

Vol. 1128: J. Elschner, Singular Ordinary Differential Operators and Pseudodifferential Equations. 200 pages. 1985.

Vol. 1129: Numerical Analysis, Lancaster 1984. Proceedings. Edited by P.R. Turner. XIV, 179 pages. 1985.

Vol. 1130: Methods in Mathematical Logic. Proceedings, 1983. Edited by C.A. Di Prisco. VII, 407 pages. 1985.

Vol. 1131: K. Sundaresan, S. Swaminathan, Geometry and Nonlinear Analysis in Banach Spaces. III, 116 pages. 1985.

Vol. 1132: Operator Algebras and their Connections with Topology and Ergodic Theory. Proceedings, 1983. Edited by H. Araki, C.C. Moore, Ş. Strătilă and C. Voiculescu. VI, 594 pages. 1985.

Vol. 1133: K.C. Kiwiel, Methods of Descent for Nondifferentiable Optimization. VI, 362 pages. 1985.

Vol. 1134: G.P. Galdi, S. Rionero, Weighted Energy Methods in Fluid Dynamics and Elasticity. VII, 126 pages. 1985.

Vol. 1135: Number Theory, New York 1983–84. Seminar. Edited by D.V. Chudnovsky, G.V. Chudnovsky, H. Cohn and M.B. Nathanson. V, 283 pages. 1985.

Vol. 1136: Quantum Probability and Applications II. Proceedings, 1984. Edited by L. Accardi and W. von Waldenfels. VI, 534 pages. 1985.

Vol. 1137: Xiao G., Surfaces fibrées en courbes de genre deux. IX, 103 pages. 1985.

Vol. 1138: A. Ocneanu, Actions of Discrete Amenable Groups on von Neumann Algebras. V, 115 pages. 1985.

Vol. 1139: Differential Geometric Methods in Mathematical Physics. Proceedings, 1983. Edited by H. D. Doebner and J. D. Hennig. VI, 337 pages. 1985.

Vol. 1140: S. Donkin, Rational Representations of Algebraic Groups. VII, 254 pages. 1985.

Vol. 1141: Recursion Theory Week. Proceedings, 1984. Edited by H.-D. Ebbinghaus, G.H. Müller and G.E. Sacks. IX, 418 pages. 1985.

Vol. 1142: Orders and their Applications. Proceedings, 1984. Edited by I. Reiner and K. W. Roggenkamp. X, 306 pages. 1985.

Vol. 1143: A. Krieg, Modular Forms on Half-Spaces of Quaternions. XIII, 203 pages. 1985.

Vol. 1144: Knot Theory and Manifolds. Proceedings, 1983. Edited by D. Rolfsen. V, 163 pages. 1985.

Vol. 1145: G. Winkler, Choquet Order and Simplices. VI, 143 1985.

Vol. 1146: Séminaire d'Algèbre Paul Dubreil et Marie-Paule M Proceedings, 1983–1984. Edité par M.-P. Malliavin. IV, 420 1985.

Vol. 1147: M. Wschebor, Surfaces Aléatoires. VII, 111 pages.

Vol. 1148: Mark A. Kon, Probability Distributions in Quantum St Mechanics. V, 121 pages. 1985.

Vol. 1149: Universal Algebra and Lattice Theory. Proceedings Edited by S. D. Comer. VI, 282 pages. 1985.

Vol. 1150: B. Kawohl, Rearrangements and Convexity of Level PDE. V, 136 pages. 1985.

Vol 1151: Ordinary and Partial Differential Equations. Proce 1984. Edited by B.D. Sleeman and R.J. Jarvis. XIV, 357 pages.

Vol. 1152: H. Widom, Asymptotic Expansions for Pseudodiff Operators on Bounded Domains. V, 150 pages. 1985.

Vol. 1153: Probability in Banach Spaces V. Proceedings, 1984. by A. Beck, R. Dudley, M. Hahn, J. Kuelbs and M. Marcus. pages. 1985.

Vol. 1154: D.S. Naidu, A.K. Rao, Singular Pertubation Anal Discrete Control Systems. IX, 195 pages. 1985.

Vol. 1155: Stability Problems for Stochastic Models. Proce 1984. Edited by V.V. Kalashnikov and V.M. Zolotarev. V pages. 1985.

Vol. 1156: Global Differential Geometry and Global Analysis Proceedings, 1984. Edited by D. Ferus, R.B. Gardner, S. He and U. Simon. V, 339 pages. 1985.

Vol. 1157: H. Levine, Classifying Immersions into $\mathbb{R}^4$ over Stable of 3-Manifolds into $\mathbb{R}^2$. V, 163 pages. 1985.

Vol. 1158: Stochastic Processes – Mathematics and Physics. P dings, 1984. Edited by S. Albeverio, Ph. Blanchard and L. Str 230 pages. 1986.

Vol. 1159: Schrödinger Operators, Como 1984. Seminar. Edite Graffi. VIII, 272 pages. 1986.

Vol. 1160: J.-C. van der Meer, The Hamiltonian Hopf Bifurcati 115 pages. 1985.

Vol. 1161: Harmonic Mappings and Minimal Immersions, Mont 1984. Seminar. Edited by E. Giusti. VII, 285 pages. 1985.

Vol. 1162: S.J.L. van Eijndhoven, J. de Graaf, Trajectory S Generalized Functions and Unbounded Operators. IV, 272 1985.

Vol. 1163: Iteration Theory and its Functional Equations. Procee 1984. Edited by R. Liedl, L. Reich and Gy. Targonski. VIII, 231 1985.

Vol. 1164: M. Meschiari, J.H. Rawnsley, S. Salamon, Ge Seminar "Luigi Bianchi" II – 1984. Edited by E. Vesentini. V pages. 1985.

Vol. 1165: Seminar on Deformations. Proceedings, 1982/84. by J. Ławrynowicz. IX, 331 pages. 1985.

Vol. 1166: Banach Spaces. Proceedings, 1984. Edited by N. and E. Saab. VI, 199 pages. 1985.

Vol. 1167: Geometry and Topology. Proceedings, 1985. Edite Alexander and J. Harer. VI, 292 pages. 1985.

Vol. 1168: S.S. Agaian, Hadamard Matrices and their Applicatio 227 pages. 1985.

Vol. 1169: W.A. Light, E.W. Cheney, Approximation Theory in T Product Spaces. VII, 157 pages. 1985.

Vol. 1170: B.S. Thomson, Real Functions. VII, 229 pages. 1985

Vol. 1171: Polynômes Orthogonaux et Applications. Procee 1984. Edité par C. Brezinski, A. Draux, A.P. Magnus, P. Maron Ronveaux. XXXVII, 584 pages. 1985.

Vol. 1172: Algebraic Topology, Göttingen 1984. Proceedings. by L. Smith. VI, 209 pages. 1985.